T0243069

CAMBRIDGE LIBRARY COLLECTION

Books of enduring scholarly value

Technology

The focus of this series is engineering, broadly construed. It covers technological innovation from a range of periods and cultures, but centres on the technological achievements of the industrial era in the West, particularly in the nineteenth century, as understood by their contemporaries. Infrastructure is one major focus, covering the building of railways and canals, bridges and tunnels, land drainage, the laying of submarine cables, and the construction of docks and lighthouses. Other key topics include developments in industrial and manufacturing fields such as mining technology, the production of iron and steel, the use of steam power, and chemical processes such as photography and textile dyes.

A Practical and Scientific Treatise on Calcareous Mortars and Cements, Artificial and Natural

Having devised an artificial cement in 1817, Louis-Joseph Vicat (1786–1861) sought to share and further the science surrounding calcareous cements. His son, Joseph Vicat, went on to found the eponymous company which became an international manufacturer of cement. This work was first published in French in 1828 and is reissued here in the English translation of 1837. Vicat addresses the subject of limes, the ingredients used to prepare mortars and cements, and how these building materials are affected by environmental conditions, such as immersion in water or exposure to damp soil and inclement weather. He also compares binding products of the time with those developed by the ancient Egyptians, Romans and Greeks. The translator, J.T. Smith, provides helpful explanatory notes and clarifies technical terms. Charles William Pasley's *Observations on Limes, Calcareous Cements, Mortars, Stuccos, and Concrete* (1838) is also reissued in this series.

Cambridge University Press has long been a pioneer in the reissuing of out-of-print titles from its own backlist, producing digital reprints of books that are still sought after by scholars and students but could not be reprinted economically using traditional technology. The Cambridge Library Collection extends this activity to a wider range of books which are still of importance to researchers and professionals, either for the source material they contain, or as landmarks in the history of their academic discipline.

Drawing from the world-renowned collections in the Cambridge University Library and other partner libraries, and guided by the advice of experts in each subject area, Cambridge University Press is using state-of-the-art scanning machines in its own Printing House to capture the content of each book selected for inclusion. The files are processed to give a consistently clear, crisp image, and the books finished to the high quality standard for which the Press is recognised around the world. The latest print-on-demand technology ensures that the books will remain available indefinitely, and that orders for single or multiple copies can quickly be supplied.

The Cambridge Library Collection brings back to life books of enduring scholarly value (including out-of-copyright works originally issued by other publishers) across a wide range of disciplines in the humanities and social sciences and in science and technology.

A Practical and Scientific Treatise on Calcareous Mortars and Cements, Artificial and Natural

Louis-Joseph Vicat
Edited and translated by
John Thomas Smith

Cambridge University Press

CAMBRIDGE
UNIVERSITY PRESS

University Printing House, Cambridge, CB2 8BS, United Kingdom

Cambridge University Press is part of the University of Cambridge.
It furthers the University's mission by disseminating knowledge in the pursuit of education, learning and research at the highest international levels of excellence.

www.cambridge.org
Information on this title: www.cambridge.org/9781108071512

This edition first published 1837
This digitally printed version 2014

ISBN 978-1-108-07151-2 Paperback

A

PRACTICAL AND SCIENTIFIC

TREATISE

ON CALCAREOUS

MORTARS AND CEMENTS,

ARTIFICIAL AND NATURAL;

CONTAINING,

DIRECTIONS FOR ASCERTAINING THE QUALITIES OF THE DIFFERENT INGREDIENTS, FOR PREPARING THEM FOR USE, AND FOR COMBINING THEM TOGETHER IN THE MOST ADVANTAGEOUS MANNER; WITH A THEORETICAL INVESTIGATION OF THEIR PROPERTIES AND MODES OF ACTION.

THE WHOLE FOUNDED UPON AN EXTENSIVE SERIES OF ORIGINAL EXPERIMENTS, WITH EXAMPLES OF THEIR PRACTICAL APPLICATION ON THE LARGE SCALE.

BY L. J. VICAT,

ENGINEER IN CHIEF OF BRIDGES AND ROADS; FORMERLY PUPIL OF THE "ECOLE POLYTECHNIQUE;" MEMBER OF THE LEGION OF HONOUR, ETC., ETC., ETC.

TRANSLATED,

WITH THE ADDITION OF EXPLANATORY NOTES, EMBRACING REMARKS UPON THE RESULTS OF VARIOUS NEW EXPERIMENTS,

BY

CAPTAIN J. T. SMITH, MADRAS ENGINEERS, F.R.S.

ASSOCIATE MEMBER OF THE CIVIL ENGINEERS INSTITUTION, LATE PRESIDENT OF THE EDINBURGH PHILOSOPHICAL SOCIETY.

LONDON:

JOHN WEALE, ARCHITECTURAL LIBRARY,

59, HIGH HOLBORN.

1837.

TO

JOHN GRANT MALCOLMSON, ESQ.,

MADRAS MEDICAL ESTABLISHMENT, M.D., F.G.S., F.R.A.S.,
ETC., ETC., ETC.,

THIS WORK IS INSCRIBED,

IN TESTIMONY OF UNFEIGNED ESTEEM,

AND IN GRATEFUL ACKNOWLEDGEMENT OF NUMEROUS AND DISINTERESTED ACTS OF REAL KINDNESS,

BY HIS OBLIGED

AND VERY SINCERE FRIEND,

J. T. SMITH.

AUTHOR'S PREFACE.

THE art of composing calcareous cements was confined, till within the last few years, to the knowledge of a small number of facts, and to the observance of certain rules long since admitted into use without examination, on the authority of Vitruvius and the architects who followed him. But the rules were almost always found to be at fault, and the facts, for want of correlativeness, were of but little aid. Could we, for instance, manufacture good mortar in France, by mixing three parts in bulk of dry pit sand, or two of river sand, with one part of slaked lime derived from a white marble of great hardness? Such, however, are the proportions of admixture, and the characteristics of good limestone pointed out by Vitruvius. It was of little importance besides to those, to whom it was impossible to procure it, that the pouzzolana of Italy and the Dutch tarras were possessed of extraordinary binding qualities; that lime eminently adapted for hydraulic works was to be found at Metz, Viviers, Nismes, &c., in France, at Lœa in Upland, at Aberthaw in England, and elsewhere. With all this information, and even adding to it the discoveries of the Swedish Baggé, and Count Chaptal, regarding the transformation of some schists, and certain ochreous clays into pouzzolanas by calcination, it was not the less necessary to work by guess in most instances, or to trust to obscure analogies for the success of the most important works. One engineer vaunted the efficacy of the powder of *well-burnt* tile, another looked upon smithy slag and iron-dross as the finest ingredients. These again, on the other hand, asserted, that such substances are destitute of energy. Lastly, this difference of opinion extended even to the manipulation of the compounds. Was the lime to be slaked with much water, or to be allowed to fall to powder after having immersed it for a few seconds? Should it be applied hot or cold? &c. Every plan had its partisans; and what doubtless appears paradoxical was, that each method too was supported by experiments and testimony, of which it was hardly possible to dispute the authenticity.

We shall leave it to the reader to appreciate a state of things like this, and to decide whether such a chaos of opinions and opposing facts could or could not make up a science,—a doctrine of calcareous cements. Perhaps it may be replied, that at the epoch of which we speak, builders had learnt to erect bridges, locks, &c., without either tarras or pouzzolana, and in countries where the lime possessed no extraordinary quality. Without denying this truth, we must remark, that most of these works have not endured, nor can continue to endure, but by frequent and expensive repairs. That on many canals it has been necessary to reconstruct a great many locks, whose side walls were in a few years found to be quite stripped of mortar. That a multitude of dikes, sluices, weirs, ("barrages,") and aqueducts, of recent construction, already exhibit all the characteristics of age, without the possibility of attributing these unexpected dilapidations to any other cause than the bad quality of the mortars or cements made use of.

These facts, known to a number of engineers, have long since attested the insufficiency of the art; and this insufficiency exhibited itself more and more, owing to the multiplicity of marine works called for by a constantly increasing commerce. It was to put an end to such a state of things, that we, in 1812, commenced our experimental researches, published in 1818. The subject, so to speak, is one of intrinsic importance, and consequently to discuss it merely is sufficient to attract public attention. We may, therefore, be allowed to say, laying aside personal vanity, that the experimental researches on lime and mortars have been the subject of serious examination by chemists, architects, and engineers. Certain theoretical points, independent of general results, have given rise to these discussions, which have themselves stimulated us on our part to new labours. Experiments in contrast with one another, undertaken in various parts of the kingdom by order of Mr. Becquey, Director-General of Public Roads and Mines, while they consolidated the necessary fundamental points established by us, enlarged the domain of facts to such an extent, that it became necessary to re-digest the whole, in order to arrange and compare them together. But by this very operation, owing to the multiplication and mutual support of the truths, they have added fresh confirmation to those which we were already possessed of. They have also enabled us to contract the scale comprehending them, both by leaving us at liberty to adopt a mode of classification before impossible, as well as by affording us the power of casting into notes a crowd of details and historic or scientific documents, useful to consult, but not indispensable to the understanding of the whole.

It is, moreover, in the nature of things to become more simple, in proportion as they approach perfection; and this is the more fortunate, as now-a-days, much more than formerly, large volumes create alarm, and are no longer read.

These explanations having been given, we are anxious here to make known the fresh obligations which we labour under to the analytic and synthetic labours of Messrs. John and Berthier, on calcareous compounds and hydraulic limes; to the researches of M. Bruyére, Inspector-General of Roads and Bridges, on the manufacture of artificial pouzzolanas and cements, resulting from the calcination of clays combined with a small proportion of lime; to the very remarkable experiments of Messrs. Avril and Girard de Caudemberg, Engineers, the first on the psammites of Finisterre—the second on the arenes of Perigord; to the examination of the limes of Russia by Col. Raucourt, an examination which has led that Engineer to enrich the science with numerous important observations; and lastly, to the interesting results obtained by M. Lacordaire, Engineer, in using the hydraulic limestones imperfectly burned, as natural cements.

There are services of another kind which require no less acknowledgment — such is the generous and enlightened manner in which M. Bruyére, Inspector-General, first, in 1818, obtained for my work the attention and support of the Director-General and Council of Roads and Bridges; such also is the succour afforded to this work by the honourable mention of it which Messrs. Gay-Lussac and Thenard have been pleased to make in their Lectures in the "Ecole Polytechnique;" in the Syllabus of his Lectures on Building by M. Sgauzin, Inspector-General; and lastly, in an able Report to the Academy of Sciences, by M. Girard, Member of the Academy.

TRANSLATOR'S PREFACE.

The merits of M. Vicat's valuable researches into the composition of mortars and cements are already too well known, to render it necessary for me to apologise for an endeavour to extend their usefulness, by submitting them to the public in a more accessible form. But as the motives which induced me to undertake this work, and have encouraged me to persevere in its fulfilment, may require explanation, I ought not to refrain from making them known, nor from claiming that indulgence for the result of my labours, which the peculiar circumstances under which they have been accomplished render necessary.

Having been occupied for many years in the construction and repairs of numerous public buildings, the charge of which devolved upon me in the performance of staff duties, I was long embarrassed, in the endeavour to give durability to works executed under my superintendence, by many difficulties arising from the defective quality of the cements employed, the dampness of the situation, and other causes at the time unknown.

Anxious to remedy these evils, I engaged in a series of experiments, in which numerous modifications of the processes previously employed, and every suggestion which could be gleaned from the scattered hints contained in the writings of the various English authors who have incidentally touched on the subject, were put to trial, both with reference to the durability of the compounds, as well as their economy on the large scale. But although these endeavours were followed by many promising results, it was not until I became possessed of M. Vicat's Work, that the theory of the composition of mortars and cements was developed in a sufficiently satisfactory and comprehensive manner, to enable me to take a full view of the varied resources found within the limits of almost every locality, for the fulfilment of the objects of which I was in search. But, systematic and plain as M. Vicat's instructions and experiments are when well understood; yet it was not without much labour, in repeating many of the experiments, and the perusal of other French authors on the same subject, that I was enabled to overcome the difficulties occasioned by my imperfect acquaintance with the exact meaning of the numerous technical terms employed in it, and fully to appreciate the originality and appropriateness of the experiments, and the depth and philosophical accuracy of the reasoning founded on them. Having surmounted these obstacles, and felt the great value of the copious information placed at my disposal, I could not look back upon the pains which it had cost me to effect my object without being led to consider, that others similarly situated with myself might have the same impediments to contend with; and that I might assist future inquirers, by placing the labours of M. Vicat within the reach of those, who might not possess sufficient leisure to give that attention to his work which I had found to be indispensable.

Of the desirableness of such a work, indeed, it needed but little consideration to satisfy me; for though intimately connected as such researches are with the success and durability of our most important constructions, and with the security and domestic comforts of every class of civilized society; it is remarkable, that since the publication of Dr. Higgins, now rendered obsolete by the rapid strides which the art has taken since his time, no English work on this subject has yet appeared. Nor have the investigations connected with it, hitherto, attracted the attention of any of the distinguished philosophers, to whom science and the arts are in other respects so largely indebted.

At a time, therefore, when the rapidly extending demands of a quickly progressive civilization, daily give birth to new and stupendous undertakings, there could be no doubt as to the benefits which must result from bringing before the public, the labours of those who have devoted themselves to the study of this very important and hitherto neglected branch of Architecture, and of placing within their reach the many valuable facts brought to light by them.

But of my own fitness for this undertaking, even with the advantages of expected leisure under which it was commenced, I could not avoid feeling the greatest diffidence; nor should I have ever ventured to incur so great a responsibility, had I not been encouraged by the consideration, that much which could not fail to be of service, might be effected by the mere exertion of persevering industry. And that, although, as I felt conscious, numbers might have been found infinitely better qualified than myself to do justice to the task; yet, that the very circumstance of superior fitness, joined to the increasing demands upon the talented members of the profession to whose province it would most properly belong, would be sufficient to prevent the public from ever deriving the benefit of their assistance.

Under the influence of these considerations, therefore, and in the hope of thereby finding useful and instructive occupation, for the leisure which an absence from my duties on account of ill health would afford me, I made up my mind to commence the task. But I had not proceeded far, when I was unfortunately deprived of the advantage upon which I had principally relied for success, by being called upon to apply the time I had intended to devote to this object, to the service of Government in a different pursuit; whereby I was deprived of the ability to devote that attention and study to it, which it was my earnest wish to have done.

These causes must, I fear, necessarily be pleaded as an excuse, for those inaccuracies which I cannot hope to have escaped from; and which will, I trust, be treated with indulgence. In the general design and execution of the work, however, I have not failed to keep in view the convenience of the reader, in so far as it lay in my power to add to it, or to the general usefulness of the volume. I have therefore, throughout, endeavoured to communicate whatever information it was in my power to collect, either from the published works of others, or from my own experiments, in illustration or support of the opinions or statements contained in the text. The whole of the measures made use of have, also, except when clearly unnecessary, been reduced to the corresponding English standard; a process which has also been applied to the very valuable results collected together in the Tables. Thus, these experiments now admit of a ready comparison with similar ones made in this country, and the reader will find no difficulty in forming a clear apprehension as to the efficiency of the processes to which they are applied as tests. It may be proper to add, that these calculations have been made from tables of the correspondence of English and French weights and measures, given at the end of Ure's Chemical Dictionary (Edition of 1824).

In the first, or more practical part of the volume, explanations have been given in the notes, of such scientific terms as may not be familiar to the general reader. This, however, has not been done in respect to all the notes in the Appendix; as many of them, consisting of purely scientific reasoning, could not, in the limited space of a note, have been rendered perfectly intelligible to those by whom the terms themselves were not understood.

These additions, with that of a copious Index, and a more distinct separation of the various subjects severally treated, will, I hope, materially assist the perusal; more particularly by obviating the confusion liable to be occasioned by the apparent contradiction of the different directions given in the work, applicable to various circumstances. In the latter part of the volume, I have ventured to take a liberty with respect to its arrangement, to which I was prompted by a desire to attract the attention of the scientific reader to a subject hitherto little noticed. I have for this purpose converted the "note on the theory of calcareous mortars and cements" into a distinct chapter (the seventeenth); to which the appendix to that note, together with the particulars of some experiments by myself in prosecution of the same subject, have formed an appropriate appendix. To this more prominent position in the body of the work, the theoretical investigations above mentioned seemed to me to be entitled, both from their close connexion with and essential influence over successful practice, as well as from their intrinsic value and philosophical interest. Its discussion is accompanied, moreover, by so many hints calculated to awaken attention and stimulate inquiry, whilst so little seems to be wanting to complete the evidence, that we may soon hope to be possessed of a sufficient number of facts, to form the basis of a correct theory of the hitherto ill-understood causes of solidification, under all the various circumstances in which it takes place.

In regard to the use of some new terms which I have found it necessary to apply, it may be right to explain, that after trying many substitutes taken from the technical language of the best authorities on this subject, I found M. Vicat's classification so different and so much more methodical than any which has hitherto obtained in this country, that I should have run the risk of sacrificing the clearness of his arrangement, had I attempted to introduce any synonymes taken from the English language to express his meaning, in lieu of the simple translation to which I have confined myself. Other words, used for the purpose of defining substances hitherto classed by us under a more general category, and consequently intended to mark distinctions at present unknown, such as "arenes," "psammites," &c., I have thought it advisable to convert at once into English terms, taking care to explain their meaning on their first application.

Of the particular merits of the numerous practical instructions and varied processes contained in this volume, it would perhaps be premature in me here to speak; the most satisfactory recommendation being the experience of those, who may have occasion to put them to the test of actual trial. I ought not, however, to omit to notice a circumstance which it would be injustice to M. Vicat not to refer to. This is, that although the processes he has laid down for the manufacture of artificial hydraulic compounds are of a comprehensive nature, capable of accommodating their results to the exact wants of the architect in every situation, thus including all the various kinds of Roman Cements, &c.; yet, it will be observed, that his own practice seems to have been chiefly confined to the adoption of the hydraulic *limes*, in lieu of the more energetic cements more generally used in this Country. This preference may expose him to the opposition of many firmly-established usages and opinions, where the latter practice has so long and successfully prevailed, with respect to the justice of which it would not become me to hazard an opinion. M. Vicat has, however, relieved me from that necessity, by expressing his own very decidedly, in his declaration (in Chapter XV., Art. 263), that the superior adhesion of the hydraulic limes over our (so called) Roman cements, must inevitably, in time, give them the preference, whenever the comparative merits of the two are fairly known and appreciated. Now, without entering upon the discussion of this question, I may remark, that it appears to be one in which a contrariety of opinion may be occasioned by a difference of situation and circumstances. Thus it may perhaps be important, in considering the merits of the two systems, to recollect, that in one the means of minute mechanical division are an essential element, in the other, that it is unnecessary; and that this element which in one situation may be obtained at a cheap rate, may in another be expensive or unattainable. The hydraulic limes, therefore, which do not require to be *ground* previous to use, are at all events, whatever may be their other merits, more especially suitable to those situations where the facilities of mechanical agency cannot be resorted to. This circumstance would in itself be sufficient to justify M. Vicat's opinion; but I have now referred to it principally to point out, that the use of *ground cements*, valuable as they are in our constructions, are better adapted to the vicinity of a large capital, where it is of little importance that the builder becomes dependent upon others for his supply, than for a remote situation or a new country, in which the *unground* limes cannot fail to be preferred, from the facility with which they may be prepared by the mason himself. The difference, in fact, consists in this, that the *ground cements*, of whatever kind, will ever be furnished by *manufacturers*, whereas the hydraulic limes may at all times be prepared by the common workman, without machinery, and at a cost not much exceeding that of common lime (vide note to App. XVIII). And it will be in reference to this advantage, in addition to those pointed out by M. Vicat, and in opposition to the inconveniences which may be occasioned by the defect peculiar to them, their comparatively tardy solidification, that the engineer will be guided in making the selection best suited to his situation and exigencies.

Moreover, it is not merely in the accuracy of the details of his valuable invention, that M. Vicat has done the most service to the profession by the publication of his work; we must not forget the variety of other processes which he has illustrated and verified by numerous and exact experiments, and by which he has increased the resources of the practical engineer in every situation. And it is by the broad light thrown upon the relations of the numerous but ill-known ingredients, that he has placed within his reach a clew for the formation of compounds, hitherto guided by empirical rules, seldom derived from and therefore not adapted to the circumstances under which they are to be applied.

It now only remains for me to express my acknowledgments for the assistance of which I have availed myself, in the execution of my small part of this volume. Of the published works to which I have had occasion to refer, I have made a point of duly stating the authority to whom I have been indebted; and to the distinguished authors to whom these obligations are due, I have merely to add the names of my friend Dr. Malcolmson, whose valuable services will be recognised in various parts of the Work, and of Colonel Sim, of the Madras Engineers, whose kindness in the ready communication of the results of his extensive experience in the processes for the manufacture of the celebrated mortars and stuccoes of Madras, has added another favour to the many debts of gratitude long due to him.

TABLE OF CONTENTS.

SECTION I.

VARIOUS LIMES, OR AGENTS OF ADHESION IN CALCAREOUS MORTARS AND CEMENTS.

CHAPTER I.

CHAPTER II.

CHAPTER III.

CHAPTER IV.

CHAPTER V.

SECTION II.

VARIOUS INGREDIENTS WHICH UNITE WITH LIME IN THE PREPARATION OF CALCAREOUS MORTARS AND CEMENTS.

CHAPTER VI.

CHAPTER VII.

CHAPTER VIII.

CHAPTER IX.

SECTION III.

COMBINATION OF THE ELEMENTS OF CALCAREOUS MORTARS AND CEMENTS.

CHAPTER X.

CHAPTER XI.

CHAPTER XII.

CHAPTER XIII.

CHAPTER XIV.

CHAPTER XV.

CHAPTER XVI.

CHAPTER XVII.

APPENDIX.

NOTES ON CHAPTER I.

NOTES ON CHAPTER II.

NOTES ON CHAPTER III.

NOTES ON CHAPTER IV.

NOTES ON CHAPTER V.

NOTES ON CHAPTER VII.

NOTES ON CHAPTER VIII.

NOTES ON CHAPTER IX.

NOTES ON CHAPTER X.

NOTES ON CHAPTER XI.

NOTES ON CHAPTER XIII.

NOTES ON CHAPTER XIV.

NOTES ON CHAPTER XV.

NOTES ON CHAPTER XVI.

NOTES ON CHAPTER XVII.

TABLES.

TABLE No. 1.

TABLE No. 2.

TABLE No. 3. (bis.)

TABLE No. 3.

TABLE No. 4.

TABLE No. 5.

TABLE No. 6.

TABLE No. 7.

TABLE No. 8.

TABLE No. 9.

DOUBLE TABLE No. 10.

TABLE No. 11.

TABLE No. 12.

TREBLE TABLE No. 13.

DOUBLE TABLE No. 14.

TABLE No. 15.

TABLE No. 16.

TABLE No. 17.

A

COMPENDIUM,

&c. &c.

SECTION I.

CHAPTER I.

OF CALCAREOUS MINERALS, AND THE VARIOUS KINDS OF LIME THEY FURNISH.

1. CALCAREOUS minerals are substances essentially composed of lime and carbonic acid;[a] they always dissolve, either wholly or in part, in weak acids, with a more or less brisk effervescence, and may be scratched with an iron point.

2. Limestones are sometimes pure, that is to say, wholly composed of lime and carbonic acid; at others, the lime is associated in intimate combination with silica, alumina,[b] magnesia, with quartz in grains, oxide of iron, manganese,[c] bitumen, or sulphuretted-

[a] Carbonic acid is a gas composed of one equivalent of carbon and two of oxygen: it is transparent and colourless, and incapable of supporting combustion or respiration: it combines with the alkalies, oxides, &c., forming the class of *carbonates*.—TR.

[b] Alumina, or oxide of aluminum as it is now termed, is the substance which forms the basis of the plastic clays.—TR.

[c] Manganese is a metal similar in appearance to iron, but rarely met with in the metallic state.—TR.

hydrogen.[d] The presence of these substances one by one, or two and two, or three and three, &c., constitutes the various kinds of limestone, which are further subdivided into different varieties.

3. Mineralogists distinguish the argillaceous, magnesian, sandy (arénacés), ferruginous, manganesian, bituminous, fetid, &c.; and then in each of these kinds point out varieties of form and structure, which they specify under the denominations of foliated, lamellar, saccharoidal,[e] granular, compact, globular, coarse,[f] chalky, pulverulent, pseudo-morphous,[g] concretionary, nodular, ("geodiques"[h]) incrusting, &c., &c. (App. I.)

[d] Sulphuretted-hydrogen is a compound gas, containing one equivalent of sulphur, and one of hydrogen. It is liable to be extricated on the decomposition of a metallic sulphuret by water, whence it is not an uncommon natural product.—Tr.

[e] "The whitest and most esteemed (granular limestone), from its resemblance to sugar, has been termed by the French mineralogists chaux carbonatée saccharoide; but it has more generally, from its important uses in the arts, obtained the name of *Statuary Marble.*" —*Phillips's Mineralogy,* edit. 1816.

[f] "The houses of Paris are built of a large-grained and soft calcareous stone, which is incapable of polish, and is of a dingy white, grey, or yellowish-white colour. It is found in immense horizontal beds, forming the plains south of Paris. It is a very impure limestone, and furnishes when calcined a very bad lime. The use to which it is put has occasioned its receiving the familiar name of Pierre à bâtir. Haüy describes it under that of chaux carbonatée *grossière.*"—*Ibid,* p. 129.

[g] "Minerals exhibiting impressions of the forms peculiar to the crystals of other substances are said to be pseudo-morphous."—*Ibid,* p. l.

[h] "A geode is a hollow ball. At Oberstein, in Saxony, are found hollow balls of agate lined with crystals of quartz or amethyst, which are termed geodes."—*Ibid,* p. xlvi.

4. It is useful to be acquainted with this nomenclature; but that which it is of the most importance for the builder to be aware of, is, that each variety of limestone furnishes a peculiar kind of lime, distinct in colour, weight, in its avidity for water, and above all by the hardness it acquires on being mixed intimately after slaking with the earthy substances known under the names of sand, puzzuolana, &c.

5. The physical characters which serve to distinguish calcareous minerals, fail to give any certain indication of the qualities of the lime they contain. Even chemical analysis itself is mostly but an approximate mode of investigating them, in addition to its being only within the reach of those familiar with laboratory manipulation. Experience by actual trial ought to be the builder's only guide. (App. II.)

6. We readily assure ourselves that a mineral belongs to the calcareous class, by trying it, as I have already said, with an iron point, and a weak acid.[1] Having established this fact, we reduce the experimental specimens to the average size of a large walnut, and fill with them a crucible ("gazette"), or any other vessel of baked earth, pierced with holes to favour the circulation of the air: we place the whole in the middle of a pottery furnace, (a brick or lime-kiln will answer equally as well, if it is heated by the flame of a fire of wood or furze,) and at the end of its calcination (fifteen to twenty hours) we remove the material, and introduce it while still warm into large-mouthed bottles, quite dry, and which we immediately

[1] Vide Articles 1 and 23.

close hermetically. The object of this precaution is to preserve the lime in all its activity (causticity), till the moment fixed on to submit it to experiment.[k]

7. When we feel disposed to begin this experiment, we remove the lime from the bottle, and take as much in bulk as would about fill a quart measure (including voids): we put it into a cloth bag of an open material, or rather a small basket; we immerse the whole for five or six seconds only in pure water; we drain it an instant, and then empty the bag or basket into a stone or cast-iron mortar.

8. The following are the different phenomena which may ensue after this immersion:—

1st. The lime hisses, decrepitates, swells, gives out a great quantity of hot vapours, and falls to powder instantaneously, or nearly so.

2nd. The lime remains inactive for a space more or less long, not exceeding five or six minutes; after which the phenomena above described manifest themselves with energy.

3rd. The lime exhibits no alteration even after five or six minutes; a quarter of an hour even may elapse before it appears to change its state. However, it begins to smoke and crack with little or no decrepitation: the vapour formed is less abundant, and not so hot, as in the preceding case.

4th. The phenomena do not commence till an hour, and sometimes till many hours, after the exmersion.

[k] In order to make sure that the calcination is complete, it would be as well to subject a small portion of the lime to trial, by slaking it with a little water, and adding dilute muriatic acid: if sufficiently burned it ought to dissolve without effervescence.—TR.

Cracks form without decrepitation, slight fumes are given out, and but little heat is disengaged.

5th. The phenomena commence at periods very variable, but are hardly sensible; the heat developed barely manifests itself except to the touch; the pulverulence is but obscurely marked, and sometimes does not ensue at all.[1]

9. In no case is it necessary to wait till the effervescence has ceased in order to finish the slaking; as soon as the disaggregation manifests itself, we pour water into the vessel, not upon the lime, but on one side, in such a manner that it may flow freely to the bottom, whence it is sucked up by those portions of the material which are farthest advanced. We stir it at the same time with a spatula; continuing to add water if it be required, but with care not to drown the mixture. Lastly, we substitute the pestle for the spatula, and work the whole up to a stiff clayey consistency.

10. Thus prepared, the lime must be left to itself till the more sluggish portions have completed their developement. The termination of this part of the operation is indicated by the perfect cooling of the whole mass. It lasts from two to three hours, and sometimes more.

[1] If the calcined mineral (having been proved to be *calcareous* by trial with dilute acid—Articles 1 and 6) should not slake at all, or very imperfectly, it must be reduced mechanically to a perfectly impalpable powder, without the addition of water, and then dealt with as afterwards explained (Articles 11 and 12). Many of the most energetic and useful of the water-cements, such as the Yorkshire, the Harwich, and Sheppy cements, require to be treated in this manner.—Tr.

11. We now take the lime again with the pestle, and add water if it be required, in such quantity as to give us a paste as stiff as possible, yet not entirely deprived of a certain degree of ductility. Its consistency may be compared to that of clay ready to be worked up in the manufacture of pottery.

12. We then take any vessel of greater height than breadth; (a china mustard-pot, or a large drinking glass, will answer the purpose very well;) we transfer the lime to it in such quantity as to fill it up about two-thirds or three-fourths, striking the bottom of it with the palm of the hand or on a block to cause the material to settle down and spread itself on the bottom of it; we then label it carefully, and immerge the whole without delay, noting the day and hour of the immersion.

13. In studying successively during fourteen years the most remarkable limes of this kingdom treated in this manner, I have been led to arrange them in five categories, distinguished by the following denominations (App. III.):—1st, Rich limes; 2nd, Poor limes; 3rd, Limes slightly hydraulic; 4th, Hydraulic limes; 5th, Limes eminently hydraulic.

14. The *rich* limes are such as may have their volume doubled, or more, by slaking in the ordinary manner,[m] and whose consistency after many years of immersion remains still the same, or nearly the same as on the first day, and which dissolve to the last grain in pure water frequently changed.[n]

[m] Vide Article 56, and App. XXII.—Tr.

[n] Solubility in water may be used as a convenient test of the

15. The poor limes are such as have their volume but little or not at all augmented by slaking, and which, in other respects, exhibit in the water very nearly the same phenomena as the rich limes, but with this difference, that they only dissolve partially, leaving a residue of no consistency.[o]

16. The moderately hydraulic limes will *set*[p] in fifteen or twenty days after immersion, and continue to harden; but their progress becomes more and more slow, particularly after the sixth to the eighth month. After one year their consistency is about equal to that of hard soap. They dissolve also in pure water, but with great difficulty; their expansion by slaking ("foisonnement"[q]) is variable; it fre-

proper calcination of rich lime: in this case, if a little coarse sugar be melted previously in the water, it will very much increase its solvent power.—Tr.

[o] Rich limes also may leave an insoluble residue (of carbonate of lime) if either insufficiently burned, or if they should have been much exposed to the air, and become partially regenerated by the absorption of carbonic acid. Should there be any reason to suspect this, a drop or two of muriatic acid should be added to the water; when, if the residue be the carbonate of lime, formed as above explained, it will dissolve with effervescence; but if it be composed of the silicious matter of a "poor lime," it will remain insoluble.—Tr.

[p] Vide Article 20.

[q] "Lorsqu'on éteint la chaux commune avec de l'eau, á sa sortie du four, pour la reduire en pâte, on trouve qu'elle augmente considerablement de volume; cette augmentation est telle qu'une partie de chaux vive mesurée en volume en produit quelque fois plus de trois mesurée à l'état de pâte épaisse: c'est ce qu'on appelle le foisonnement."—*General Treussart, Memoire sur les Mortiers Hydrauliques*, p. 3.

quently reaches the limits of the poor limes, without ever attaining that of the rich limes.

17. The hydraulic limes *set* after six or eight days' immersion, and continue to harden. The progress of this induration may be continued to the twelfth month, although the greatest part of the effect will be attained after six months. At this period, the hardness of the lime may be already compared with that of the very soft kinds of stone, and the water ceases to have any action on it. Their expansion by slaking is always small, like the poor limes.

18. The eminently hydraulic limes *set* from the second to the fourth day of immersion. After one month they are already very hard, and altogether insoluble. At the sixth month they appear like the absorbent calcareous stones, whose surface admits of being cut. They splinter under a blow, and present a slaty fracture. Their expansion by slaking is constantly small, like the poor limes.

19. In other respects, the rich, and poor, and hydraulic limes of all grades, may be white, grey, mouse-coloured, red, &c.[r] (App. IV.)

[r] As no reference is here made to the properties of magnesian limestones, it may be useful to describe a mode by which they may be recognised. Though capable of setting under water, they may be entirely soluble in dilute acid; but magnesia not combining with water till exposed for some time in contact with it, the mineral, if containing a large proportion of this substance, will either not slake at all, or will do so imperfectly, and will gain much less in weight than an equal quantity of calcined rich lime would do. Upon this property of magnesian limestone, Mr. Prinsep has founded a very neat and simple process for estimating the proportions of the consti-

20. We say that a lime has *set*, when it bears without depression a knitting-needle of 0.12 cent. (.047 or nearly $\frac{1}{20}$ inch) diameter, filed square at its extremity, and loaded with a weight of 0.30 kil. (about 10 ozs. 9 drs. avoirdupois weight). In this state the lime will resist the finger pushed with the mean strength of the arm, and it is incapable of altering its form without fracture.

21. The chemical examination of the minerals which supply the various kinds of lime of the preceding categories, points out, in a general way, as follows:—

1st. As furnishing the rich limes: 1st, the pure limestones, or such as contain only an admixture of from .01 to .06 of silica, alumina, magnesia, iron, &c., taken separately, or two and two, three and three, &c.; 2dly, the simple, bituminous, or fetid limestones.[8]

2nd. Such as form the poor limes: 1st, limestones associated with silica in the state of sand, magnesia, the oxides of iron and manganese, in variable proportions, but limited to .15 to .30 of the whole, whether these principles exhibit themselves one by one, two and two, three and three, or all together.

3rd. As furnishing the slightly hydraulic limes:

tuent ingredients, of which an explanation will be found in the notes. (Vide App. V.)—Tr.

[8] "Swine-stone, or stink-stone, so called from the strong fetid odour given out when scraped or rubbed, is found massive and compact, and of various shades of grey, brown, and black. By calcination it becomes white, and burns into quicklime. The offensive odour which it gives out when scraped is considered to be owing to its including sulphuretted-hydrogen: it is commonly attributed to bitumen, which does not seem to enter into the composition of swine-stone."—*Phillips's Mineralogy*, p. 130.

the limestones united with clay, magnesia, iron, and manganese, the relative proportions of which being variable, but not exceeding .08 to .12 of the whole. The oxides of iron and manganese also, being present either in the relations one to one, two and two, three and three, &c., &c., or even entirely wanting.

4th. The ingredients constituting the hydraulic limes. These are the limestones containing silica, alumina, magnesia, iron, and manganese, in variable relative proportions, amounting to not more than from fifteen to eighteen hundredths of the whole, and in other respects such, that the silica always has a predominance, no matter whether the other substances appear, as one to one, or two and two, &c., &c. The iron, manganese, and magnesia, may also be entirely wanting.

5th. The minerals affording the eminently hydraulic limes are, the limestones which contain silica, alumina, magnesia, iron, and manganese, in different relative proportions, but usually limited to from twenty to twenty-five hundredths of the whole; the silica always predominating, sometimes to the extent of forming of itself more than half the whole; and the other substances only occurring one by one, two and two, or three and three, &c. It is very seldom that we meet with them all at one time. The magnesia, and still more, the manganese, are very frequently absent.

22. In the present state of our knowledge regarding the different varieties of lime, it is impossible to say whether there exist certain determinate proportions of silica alone, or of silica and alumina, or of silica and magnesia, &c., &c., which, by their intimate association with the same quantity of calcareous

matter, are capable of producing limestones of equal energy. But one thing which is certain, and which it is important to recollect, is, that no perfectly hydraulic mortar exists without silica,[t] and that all lime which can be so denominated, is found by chemical analysis to contain a certain quantity of clay, made up of silica and alumina in proportions similar to what constitutes the ordinary clays. (App. VII.)

23. The preceding observations being clearly understood, we see how easy it is in a few minutes to detect a mineral composed of rich lime, by dissolving three or four grammes (fifty or sixty grains) of it in dilute nitric or muriatic acid. If there should be left no insoluble residue, or if that residue be very trifling, it is useless to carry the experiment any farther.[u] In the other case, in order to classify the mineral, it is necessary to reduce it into lime, and proceed in the manner above described.[v] The time which such lime takes to *set*, if it be measured very exactly, will always be sufficient to afford an approximate indication of its place.

24. Twenty years ago we knew hardly a dozen

[t] An exception must be made in favour of the Dolomites, with the properties of which Mr. Vicat does not appear to have been acquainted. Colonel Pasley found that the carbonate of magnesia itself furnished an excellent water-cement; and in some experiments, made on a small scale, I found that a paste of equal parts of the hydrates of lime and magnesia calcined together set with great firmness in a few hours, whence I infer the probability that some of the magnesian limestones may be found to be useful in this way; but I have as yet had no opportunity of making a direct experiment on this subject. (Vide App. VI.)—Tr.

[u] Vide Note to Article 19. [v] Vide Articles 9, 10, 11, and 12.

localities in France affording hydraulic lime. Now we are no longer able to reckon them.[w] Wheresoever we have been sent to look for it, it has been met with, even in Brittany. The Departments of Lot, Lot and Garronne, Tarn, Dordogne, Charente, Cher, Allier, Nièvre, the Yonne, Côte-d'Or, Ain, Isere, Jura, Doubs, Upper Rhine, &c., &c., are abundantly supplied with it.

[w] As an exemplification of the practical usefulness of the methods described in this work, for recognising with facility the minerals capable of being applied to hydraulic purposes, I may state, that whilst superintending the Department of Public Works in the Northern Division of the Madras Presidency at Masulipatam, they were the means of discovering an excellent water-lime superior to the Aberthaw in setting power, which existed in abundance in the immediate neighbourhood, but whose properties had been previously altogether unknown. And there can be no doubt that these valuable limes will be found plentifully in many other situations, when Mr. Vicat's simple methods of recognising and estimating their qualities are more generally understood. With this object the class of Indian calcareous minerals called "cancars" seems to be particularly worthy of experiment, some of them which have been already analyzed exhibiting nearly the exact proportions constituting the best cements, and others being alloyed in various ratios, corresponding with the composition of water-limes of different qualities. Of these a valuable catalogue has been published by Mr. Prinsep; but it is impossible to predict from it the value of the minerals it describes, as cements, until the exact nature of the alloy in each be determined.—Tr.

CHAPTER II.

ON THE BURNING OF LIMESTONE IN THE LARGE WAY.

25. LIMESTONE becomes lime on being deprived of its carbonic acid, and of the water it contained, whether hygrometrically or in combination. The agent employed to effect this is heat. (App. VIII.)

26. With the same heat, the calcination is effected with more ease and rapidity in proportion as the stone is of a less compact texture, to the smallness in bulk of the fragments into which it is reduced, and to its being impregnated with a certain degree of humidity. (App. IX.)

27. The contact of the air is not indispensable, but it exercises a useful influence, especially in regard to argillaceous limestone. Moreover, no limestone can be converted into lime in a vessel so close as to render the escape of the carbonic acid impossible. (App. X.)

28. Limestone which is pure, or nearly so, supports a white heat without inconvenience;[a] the compound limestone, on the other hand, alloyed in the proportions necessary to form hydraulic or eminently hydraulic lime, fuses easily. Its calcination demands certain precautions: the heat ought never to be

[a] Under the intense heat of the hydro-oxygen blowpipe this substance affords the brilliant light, the beautiful application of which to the microscope is now so well known.—TR.

pushed beyond the common red heat, the intensity being made up for by its duration.

29. The compound limestone, when too much burnt, is heavy, compact, dark-coloured, covered with a kind of enamel, especially about the angular parts; it slakes with great difficulty, and givés a lime carbonized and without energy; sometimes it will not slake at all, but becomes reduced, after some days exposure to the air, to a harsh powder altogether inert.[b]

30. The pure and compound limestones when insufficiently burnt, either refuse to slake, or slake only partially, leaving a solid kernel, a kind of sub-carbonate[c] with excess of base, which possesses properties of which I shall have to speak elsewhere. (App. XI.)

[b] This is called *dead* lime, and should be carefully distinguished from the sub-carbonates, or unburnt limes, mentioned in the next paragraph. Mr. Andrews (Mechanics' Magazine, No. 126) notices four cases in which dead lime may be formed: 1st, when the limestone contains much alumina, and it has been heated so as to harden the mass; 2nd, when the limestone contains silica, and it has been strongly heated after the expulsion of the carbonic acid; 3rd, when the limestone is so strongly heated as to be in some measure melted, in which case the carbonic acid is frequently not entirely separated from the central parts of the large lumps, and they of course effervesce with acids; 4th, when the limestone is strongly heated after the total separation of the carbonic acid, so that the lime neither splits when water is added to it, nor does it effervesce with acids.—Tr.

[c] A sub-carbonate is a combination of one equivalent of any alkali with *less* than a full equivalent of carbonic acid. Thus a combination of one equivalent of lime, with half an equivalent of carbonic acid (if such a compound could be formed), would be a sub-carbonate of lime. The meaning of the term equivalent, is the definite proportion of acid required to *neutralise* exactly a given quantity of alkali, or of alkali to neutralise a given quantity of acid.

31. The calcining of calcareous minerals constitutes the art of the lime-burner. According to situation, either firewood, faggots, brushwood, turf, or coal is used.

32. It would be tedious, as well as useless, to describe in this place all the limekilns which have been suggested or tried within the last few years. I shall content myself with saying, that the forms of interior most generally adopted are, 1st, the upright rectangular prism (pl. I. figs. 1 and 2); 2nd, the cylinder (pl. I. figs. 3 and 4); 3rd, the cylinder surmounted by an erect cone slightly truncated (pl. I. fig. 5); 4th, a truncated inverted cone (pl. I. figs. 7 and 8); 5th, an ellipsoid of revolution variously curvated, or egg-shaped kiln. (Pl. I. figs. 7, 8, 11, and 12.)

33. The rectangular kilns are in use in Nivernais, and in the south of France: they burn in them, at the same time, limestone and bricks. The limestone occupies very nearly the lower half of its capacity. The upper is filled with bricks, or tiles, laid and packed edgeways.

34. The cylindric kilns are principally employed upon works which consume a large quantity of lime in a short time. They are termed "field-kilns" ("fours de campagne"); their construction is expeditious and economical, but precarious. Above a pointed ("ogive") or oven-shaped vault, they raise, in the form of a tower, a high stack of limestone, which they enclose by a curtain of rammed earth, and sup-

Thus 22 grains of carbonic acid combine with, and exactly neutralise 28 grains of lime; the equivalents of carbonic acid and lime, therefore, are always in the proportion to one another of 22 : 28.—Tr.

port outwardly by a coarse wattling, in which care is taken to leave an opening to introduce the fire beneath the vault.

35. The kilns of the third kind are constructed in a solid and durable manner, like the four-sided kilns: no bricks are burnt in these; the largest stones occupy the lower part of the cylinder, the smaller pieces and fragments are thrown into the cone which surmounts it.

36. The kilns of the fourth and fifth kind are specially intended for burning with coal.

37. The interior wall of the kiln is generally built with bricks, or other material unalterable by heat, cemented throughout a thickness of from thirty-two to forty centimetres (12 to 15 inches nearly), with a mixture of sand and refractory clay beaten together.

38. In the flare-kilns fed by logs or brushwood, the charge always rests upon one or two vaults built up dry with the materials of the charge itself. Underneath these vaults they light a small fire, which they gradually increase as they retire, in proportion as the draught establishes itself, and gains force. On reaching the exterior they adjust the aperture at the eye of the kiln suitably, and then keep it constantly filled with the combustible. The air which rushes in, carries the flame to a distance over every point of the vaults: it insinuates itself by the joints, and is not long in extending the incandescence by degrees to the highest parts.

39. There are some kinds of stone which the fire, however well regulated, seizes suddenly, and causes to fly with detonation: we cannot, without running

the risk of spoiling the charge, use these for the construction of the vaults and piers in loading the kiln. In such a case, we employ for that purpose materials which are free from such an inconvenience.

40. Practice can alone indicate the time proper for the calcination. It varies with a multitude of circumstances, such as the more or less green, more or less dry quality of the wood; the direction of the wind, if it favour the draft, or otherwise, &c.[d] The master burners usually judge by the general settling of the charge, which varies from $\frac{1}{5}$ to $\frac{1}{6}$. In a kiln of the capacity of from 60 to 75 cubic metres,[e] the fire lasts from 100 to 150 hours: every cubic metre of lime consumes (on an average) 1.66 steres in fire-wood, 22.00 steres in faggots, and 30 steres in fascines, furze, or other brushwood.[f] (App. XII.)

41. In the coal kilns by slow heat, the stone and coal are mixed. Of all the methods of burning lime, this is certainly the most precarious and difficult; more especially when applied to the argillaceous lime-stone. A mere change in the duration or intensity of the wind, any dilapidation of the interior wall of

[d] General Treussart remarks concerning the burning of the Obernai lime, that being calcined with wood, it is very difficult to obtain a homogeneous result; there are always lumps of lime which are over-burnt, and others which are not sufficiently so. The lime-burners, he adds, ought to keep up the calcination of their lime for a *longer* period, but with a *less intense* heat; this would not consume more wood, and a much better result would be obtained.—Tr.

[e] From 211.8 to 264.75 cubic feet.—Tr.

[f] A stere is equal to one cubic metre, or 35.3 English cubic feet.—Tr.

the kiln, a too great inequality in the size of the fragments, are so many causes which may retard or accelerate the draft, and occasion irregular movements in the descent of the materials, which become locked together, form a vault, and precipitate at one time the coal, and another the stone, upon the same point; hence an excess or deficiency in the calcination.

42. Sometimes a kiln works perfectly well for many weeks, and then all at once gets out of order without any visible cause. A mere change in the quality of the coal is sufficient to lead the most experienced lime-burner into error. In a word, the calcination by means of coal, and the slow heat, is an affair of cautious investigation and habit. (App. XIII.)

43. The capacity of a furnace contributes, no less than does its form, to an equable and proper calcination. There are limits beyond which we cannot enlarge it without serious evils. We have drawn and marked in plate I., figs. 9, 10, 11, and 12, the sections of four kilns executed and tried in the Department of the Maine and Loire, by Messrs. Ollivier Brothers, manufacturers of hydraulic limes. Number 9 has been abandoned in consequence of producing lime always too much or too little burnt; number 10 answered tolerably; numbers 11 and 12 answered perfectly. (App. XIV.)

44. The bulk of coal burnt to produce a cubic metre of lime, necessarily varies with the hardness of the limestone used; but within narrow limits. When not referring to the chalks, or friable marles, we calculate on an average upon three cubic metres of lime from one of coal.

45. The calcination of limestones presents other important problems for solution; but the reasonings of the closet, unaided by experiment, will always be insufficient in a matter of this kind. It is for this reason, that we here abstain from the discussion of a number of projects more or less ingenious, but wholly theoretical. We must beg the reader nevertheless to refer to the notes; he will there find some details useful to be known, when he is called upon to apply himself specially to the burning of lime. (App. XV.)

CHAPTER III.

ON ARTIFICIAL HYDRAULIC LIMES.

46. Six years have elapsed since the publication of my first researches into this subject; and already the artificial limes have been applied to a number of important works. The canals of Saint Martin and Saint Maur made almost exclusive use of them. Nearly a thousand cubic metres[a] have been employed within five years at the harbour of Toulon. These limes have served for the fabrication of the beton[b] for the foundations of several bridges; and their consumption is increasing daily in Paris and its environs. (App. XVI.)

47. We have no longer, therefore, to attend to laboratory experiments, but indeed to a new art, very nearly arrived at perfection.

[a] Equal to 1307 cubic yards 11 cubic feet English.—Tr.

[b] I had originally translated this word into our English term "concrete;" but upon further consideration, it occurred to me, that that term is applied by us to a substance so essentially different in its nature and mode of preparation, although similar in its uses, that it would be best to avoid confusion, by a distinctive appellation. Beton, then, is a mass composed of hydraulic lime and rubble, in which the lime is usually slaked *previous* to its mixture with the other ingredients, and the mass *sets under water*. In concrete, the lime slakes *after* mixture with the rubble, and the mass cannot be immerged till after consolidation.—Tr.

48. The artificial hydraulic limes are prepared by two methods: the most perfect, but also the most expensive, consists in mixing with rich lime slaked in any way, a certain proportion of clay, and calcining the mixture; this is termed artificial lime *twice kilned.*

49. By the second process, we substitute for the lime any very soft calcareous substance (such, for example, as chalk, or the tufas), which it is easy to bruise and reduce to a paste with water. From this a great saving is derived, but at the same time an artificial lime perhaps of not quite so excellent a quality as by the first process, in consequence of the rather less perfect amalgamation of the mixture. In fact, it is impossible, by mere mechanical agency, to reduce calcareous substances to the same degree of fineness as slaked lime. Nevertheless, this second process is the more generally followed, and the results to which it leads become more and more satisfactory.[c] (App. XVII.)

50. We see that by being able to regulate the proportions, we can also give to the factitious lime whatever degree of energy we please, and cause it at pleasure to equal or surpass the natural hydraulic limes.

51. We usually take twenty parts of dry clay, to eighty parts of very rich lime, or to one hundred and forty of carbonate of lime.[d] But if the lime or its car-

[c] "The artificial hydraulic limes manufactured at Meudon under my direction, by Messrs. Brian and Saint Leger, gained the gold medal at the exhibition of the products of the useful arts in 1827." —*Original note.*

[d] The 80 parts of lime here mentioned, refer to the lime in the unslaked condition, and the 140 parts to the uncalcined mineral.

bonate should already be at all mixed (with clay, TR.) in the natural state, then fifteen parts of clay will be sufficient. Moreover, it is proper to determine the proportions for every locality. In fact, all clays do not resemble one another to such an extent as to admit of their being considered as identical: the finest and softest are the best.

52. There is at Meudon, near Paris, a manufactory of artificial lime set on foot by Messrs. Brian and Saint Leger. The materials made use of are, the

If the lime be slaked, the proportion should be increased to 110 parts. The mixture here described, is such as to produce the hydraulic limes, whose properties are similar to the Aberthaw, the analysis of which, by Mr. Phillips (Annals Phil., new series, vol. viii. p. 72), shows it to correspond nearly with the proportions here recommended; as it consists of 86.2 of carbonate of lime to 11.2 clay, (with 2.6 water and carbonaceous matter,) being at the rate of 18.2 parts clay to 140 of the carbonate of lime. The cements now commonly in use in England, are much quicker setting than these, and differ from them in being *unslaked*. They contain a greater proportion of clay, but may be manufactured artificially with equal ease, by combining such relative quantities of chalk, or lime, and clay, as will suit the purpose intended. Parker's Patent Cement, as analyzed by Sir Humphry Davy (Ure's Dictionary, art. Cement), contains 45 per cent. of clay to 55 carbonate of lime. The Yorkshire cement, 34 clay to 62 carbonate of lime. The Sheppy, 32 clay to 66 carbonate of lime. And the Harwich, which is a quicker-setting cement, 47 clay to 49 carbonate of lime. (Practical Remarks on Cements, p. 32.) These facts may serve as a guide towards the admixture of ingredients, for the formation of a compound suited to our purpose, in any situation; but for the exact proportions, recourse must be had to experiment in *every case* when new materials are to be employed. In fact, so different may be the chemical properties of apparently similar materials, that no results, however definite, or however successful in one locality, can

chalk of the country, and the clay of Vaugirard,[e] which is previously broken up into lumps of the size of one's fist. A millstone set up edgeways, and a strong wheel with spokes and felloes, firmly attached to a set of harrows and rakes, are set in movement by a two-horse gin, in a circular basin of about two metres (six feet and a half English) radius. In the middle of the basin is a pillar of masonry, on which turns the vertical arbor to which the whole system is fixed: into this basin, to which water is conveyed by means of a cock, they throw successively four measures of chalk, and one measure of clay. After an hour and a half working, they obtain about 1.50 metres cube (nearly fifty-three cubic feet English) of a thin pulp, which they draw off by means of a conduit, pierced horizontally on a level with the bottom of the basin.

53. The fluid descends by its own weight; first into one excavation, then into a second, then a third,

with safety be trusted to in another, as more than a clew for the direction of similar experiments, applied to the new materials at our disposal. For such trials the process of manufacture, and the choice of clay, &c., are explained in the text; and it is therefore merely necessary to add, that when aiming at the production of a compound similar to Roman Cement, it would be necessary to increase the dose of clay to about that indicated by the composition of the materials from which our cements are manufactured, and in which it varies, as above shown, from one-third to one-half of the whole. Particular attention should be paid to the perfect amalgamation of the materials; and the degree of calcination best suited to it should be carefully observed, before attempting to imitate the process on the large scale.—Tr.

e "A hundred parts of this clay consist of silica 63, alumina 28, oxide of iron 7, loss 2."—*General Treussart*, p. 65.

and so on to a fourth or fifth. These excavations communicate with one another at top. When the first is full, the fresh liquid, as it arrives, as well as the supernatant fluid, flow over into the second excavation; from the second into the third, and so on to the last, the clear water from which drains off into a cesspool. Other excavations, cut in steps like the preceding, serve to receive the fresh products of the work, whilst the material in the first series acquires the consistency necessary for moulding. The smaller the depth of the pans in relation to their superficies, the sooner is the above-mentioned consistency obtained.

54. The mass is now subdivided into solids of a regular form by means of a mould. This operation is executed with rapidity. A moulder, working by the piece, makes on an average five thousand prisms a day, which will measure about six cubic metres (211.8 cubic feet English). These prisms are arranged on drying shelves, where in a short time they acquire the degree of desiccation and hardness proper for calcination.[f] This may be effected by any one of the methods described in the preceding chapter. At Paris they employ a mixture of coke and coal, and the common mode of burning by slow heat, rendered necessary by that kind of combustible.

55. The artificial hydraulic limes are intended to supply the place of the natural ones in those coun-

[f] "It is necessary that the prisms be thoroughly dried previous to calcination, as experience shows, that if subjected to heat while retaining any moisture, it may deprive them almost, if not entirely, of their hydraulic properties."—*Traité sur l'Art de Faire de bons Mortiers, &c., par le Colonel Raucourt de Charleville*, p. 39.

tries where the argillaceous limestone is entirely wanting. They are sold at Paris, at from seventy to seventy-four francs per cubic metre.[g] In the country, they may be had for forty francs, on an average, when twice kilned, and will not come to more than thirty francs, when they result from a mere mixture of chalk and clay. (App. XVIII.)

[g] At the exchange of 25 francs for a pound sterling, this would be at the rate of from £2. 2*s.* 10*d.* to £2. 5*s.* 3*d.* per English cubic yard.—Tr.

CHAPTER IV.

THE SLAKING OF LIME.

First Method.

56. Quick-lime, taken as it leaves the kiln, and thrown into a proper quantity of water, splits with noise, puffs up, produces a large disengagement of hot slightly-caustic vapour,[a] (these phenomena may be more or less marked, App. XIX.) and falls into a thick paste: in this state it is termed indifferently ("*chaux fondue*," "*chaux coulee*," "*chaux amortie*") slaked lime.

57. This method is generally adopted, but they abuse it strangely. They reduce the lime to a milky consistency in a separate basin, whence it runs off into a large trench. Thus drowned, it loses the greater part of its binding qualities.

58. The rich limes, when slaked and brought to a very thick pulp, give from two to three volumes for one. The poor limes, most of the hydraulic limes,

[a] The specific gravity of pure lime is 3.08. "So much heat is produced (in slaking it) that part of the water flies off in vapour. If the quantity of lime slaked be great, the heat produced is sufficient to set fire to combustibles. In this manner vessels loaded with lime have sometimes been burnt. When great quantities of lime are slaked in a dark place, not only heat but light also is emitted, as Mr. Pelletier has observed."—*Thomson's Chemistry*, vol. i. pp. 435-6.

and all the eminently hydraulic limes, do not give, under the same circumstances, more than from one to one and a quarter, or one and a half at most.

59. Rich lime, at the moment of being quenched with much water, sometimes slakes to dryness in certain parts of the vessel, to which the water reaches in but small quantity. If we now suddenly throw a fresh supply of water upon the parts which slake in this manner, it causes a hissing resembling the immersion of a red-hot iron in water; and what is remarkable, the lime being numbed by this sudden aspersion, afterwards falls to powder very imperfectly, and continues gritty. The colder the water thrown upon it, the more marked is the effect, more especially with the rich limes. When we wish to procure a slaked lime of great fineness, (for whitewashing walls for instance,) we should have a sufficient quantity of water at first, to avoid the necessity of replenishing it at the moment of effervescence: or rather, we should supply it insensibly round the dry pieces, which will take it up spontaneously by aspiration. (App. XX.)

60. All lime becomes effete, or difficult to slake, after it has been acted on by the air: this fact is more especially remarkable in the hydraulic limes, which finally are brought to such a state, that they fall to pieces in water, without exhibiting anything but a slight disengagement of heat.

Second Method.

61. Quick-lime, plunged into water for a few seconds, and withdrawn before it commences to slake, hisses,

splits with noise, gives out hot vapours, (these phenomena may be more or less marked,) and falls to powder. It is then called slaked lime by *immersion*. It may be kept a long time in this state, provided that we shelter it from moisture. It does not again become heated on tempering it. (App. XXI.)

62. One hundred parts of rich lime slaked in this manner, do not contain, on an average, more than 18 parts of water, whilst the hydraulic limes take from 20 to 35. This fact presents itself in opposition to the phenomena exhibited by the ordinary mode of slaking.

63. The very rich limes, if we content ourselves with breaking them roughly into lumps previous to immersion, and leaving them on the ground to slake, fall with difficulty to a very fine powder. More than half is still left in solid lumps of the size of a pea, and these, when once chilled, may remain in water a long time without falling in it. This difficulty may be avoided, by reducing the fragments of quick-lime, before immerging them, to the size of a walnut,[b] and still more, by heaping them together immediately after immersion in casks or large bins: the heat is then concentrated; a great portion of the water, converted into vapour on the first instant, being unable to escape is taken up by the lime, which is thus brought to a satisfactory state of division.

64. One volume of caustic rich lime measured in powder, barely gives more than 1.50 to 1.70 of slaked powder uncompressed.

[b] "It is found by experience, that one day's labour is sufficient for bruising and immersing four-and-twenty cubic feet of quick-lime." —*Colonel Raucourt*, p. 166.

65. The hydraulic limes, under the same circumstances, give from 1.80 to 2.18.

Third Method.

66. Quick-lime, subjected to the slow and continued action of the atmosphere, is reduced to a very fine powder. During this natural slaking, there is a slight disengagement of heat, but unaccompanied by any sensible vapours.

67. The rich limes increase $\frac{2}{5}$ths in weight, and give a volume of 3.52 to one (of the quick-lime in powder). The hydraulic limes take, on an average, but $\frac{1}{8}$th of water, and give a bulk of from 1.75 to 2.55 (the powders are measured without compression, "tassement"): to obtain these results, it is necessary to seize the moment when the pulverization is complete, and by no means to operate in too damp an atmosphere.

68. Of the three processes, the ordinary mode of extinction is that which most perfectly divides the rich limes, and the hydraulic limes of all degrees,[c] and consequently, which raises their expansion to the highest

[c] Even with rich limes it appears, that the spontaneous extinction does not effect a perfect division; so that, should it be thought advantageous to adopt that method, in order to increase the binding quality of the lime, it would be desirable to obviate this defect, and improve its qualities by mechanical subdivision. General Treussart makes the following observation, in reference to common (not hydraulic) lime: "I shall add, that all the mortars made with lime slaked in powder (by immersion) were very homogeneous, whereas those which were slaked by the air, exhibited in their interior a multitude of white specks, which appeared to me to

limit. Next in order in the same respects, the spontaneous mode of extinction is more suitable to the rich, than the hydraulic, and eminently hydraulic limes; but *vice versâ* in slaking by immersion. (App. XXII.)

69. From these differences it ensues, that three equal volumes of lime, in paste of the same consistency, but slaked by different processes, contain neither the same quantity of lime, nor the same quantity of water. (App. XXIII.)

70. Every kind of lime, if exposed in its caustic state to the contact of the air in a sheltered situation, insensibly re-absorbs as much carbonic acid as is necessary to saturate it[d]: the time required for this change varies with the nature and the volume of the lime. Is it rich? ten months suffice, when we spread it in beds of only $0^{m}.02^{c}$ (rather more than three-quarters of an inch) thick. After that lapse of time, one hundred and ninety-one parts of it will be composed as follows:—

Caustic lime..........................	100
Carbonic acid	74
Water	17

be particles of lime which had absorbed carbonic acid. This was especially remarkable on breaking the mortars." It is unnecessary to point out how much this must have impaired the energy of these mortars.—TR.

[d] The quantity of carbonic acid regained from the atmosphere, *never* amounts to the full saturating dose: and as it appears to depend upon a previous re-union with *water* derived from the atmosphere, (vide note to Art. 75,) it is most probable that the time of such absorption will be regulated essentially by the state of the wea-

Is it hydraulic? In that case, under the same circumstances, the change is complete after the seventh or eighth month. At that period, 169 parts of it will contain as follows: caustic lime combined with one-fifth clay 100, carbonic acid 54, water 15.

71. All lime, when first slaked by immersion, and then exposed to the contact of the air in a sheltered situation, becomes gradually loaded with carbonic acid and water, but only up to a certain point: the amount of this change, as well as the time required for it, varies with the quality of the lime.

72. One hundred and sixty parts of rich lime so slaked, contain, after seven and a half months of exposure (as before said), caustic lime 100, carbonic acid 36.15, water 23.85.

73. One hundred and sixty-nine parts of hydraulic lime, under the same circumstances, contain, caustic lime with one-fifth of clay 100, carbonic acid 44, water 25.

74. After the period alluded to, the weight of the powders does not sensibly increase further; the balance will merely detect hygrometric changes, sometimes positive, sometimes negative. Water, however, in which we may mix these powders, dissolves a small quantity.

75. The evident consequence of these last results is, that a sudden immersion robs the rich and hydraulic limes for ever of the faculty of regaining, by a long

ther, and the humidity or otherwise of the climate in which the experiment is made.—Tr.

exposure to the air, the quantity of carbonic acid of which they have been deprived by calcination.[e]

76. In the work-yards, rich limes slaked by the "ordinary" process, are preserved by placing them in trenches nearly impermeable, and covering them over with thirty to forty centimetres (11.8 to 15.7 inches) of sand or fresh earth. (App. XXIV.) When slaked by immersion, or spontaneously, they may be kept without change for a tolerably long time, either in casks, or under sheds, in large bins covered with cloths, or with straw. (App. XXV.)

77. The hydraulic limes harden in a short time in a trench (App. XXVI): they cannot be kept long, nor especially be much carried about, without very sensible alteration, unless they be slaked by immersion, and then secured in that state in casks, or sacks of cloth. (App. XXVII.) One may, however, keep a pretty large quantity in a caustic state for five or six months, by adopting the following method.

78. We spread a layer of fifteen to twenty centi-

[e] It seems to be essential to the re-union of carbonic acid with the lime, that the latter should have previously combined with its equivalent of water; for I found that stuccoes which had been hastily formed with unslaked lime, got up with as little water as possible, were, after a couple of years' exposure, still in a caustic state quite close to the surface; while a similar stucco, composed with well-tempered lime, would be neutralized by the carbonic acid of the air to the depth of half or three-quarters of an inch in the same circumstances. This observation is also confirmed by an experiment detailed by Dr. Henry (System of Chemistry, vol. i. p. 610); for he states, that if a piece of dry quick-lime be passed into a jar of carbonic acid over mercury, no absorption whatever takes place.—TR.

metres (5.9 to 7.9 inches) thick, of the lime reduced to powder by immersion, on the floor of the shed where the supply is to be laid up: on this layer we pile up the quick-lime, packing it as close as possible. If there be no planking, the sides of the heap are finished in slopes, which are covered by a final layer of lime, taken at the moment of its undergoing immersion; this, falling to powder, lodges itself amongst the interstices of the lumps of lime, and covers it sufficiently to defend it from the air and all damp. (App. XXVIII.)

CHAPTER V.

OF THE HYDRATES[a] OF LIME (APP. XXIX), OR THE SOLIDS RESULTING FROM THE SIMPLE COMBINATION OF WATER AND LIME.

79. The study of these substances would be attended with but little interest, if it were not linked in its results to facts most important in the history of mortars.

80. Numerous preliminary trials have shown, that the quantity of water employed in slaking the lime exerts a powerful influence on the hardness of the hydrate which results. And this is easily understood. Too little water fails to bind the mass; an excess swells out the stuff, which remains light, porous, and friable, if it does not shrink proportionably in drying. Plaster[b] mixed thin, or stiff, exhibits a remarkable instance of this fact.

81. The pasty consistency which produces the greatest hardness, is at once ductile and firm. We have given it the name of "clayey," because, in fact, we can compare it to nothing better than clay which is in a state of readiness for the manufacture of pottery. It is this kind of consistency which we con-

[a] A hydrate is a compound of water chemically united with any substance. The hydrate of lime consists of about twenty-eight parts by weight of lime to nine of water. This compound is formed by the process of slaking.—Tr.

[b] Plaster of Paris is here meant.—Tr.

stantly gave to the pastes of the hydrates made use of in our experiments.[c]

Action of the Air on Hydrates.

82. Different limes prepared in this manner, and exposed in the form of quadrangular prisms to the action of the air during one year, have furnished the following results:—

1st. The carbonic acid contained in the atmosphere attacks the hydrates, fixes itself in them insensibly, and carbonises them; its influence extending from the surfaces towards the centre. The thickness of the coats thus carbonised, amounts after a year to not more than six millimetres (.236 inches, Tr.) for hydraulic limes, and to two or three (.078 inches to .118 inches, Tr.) for the rich limes. We may convince ourselves of this, by making sections of the prisms in different directions by means of a small spring saw. The coloured limes exhibit a band enveloping them, which is distinguished from the interior by a deeper tint. This tint is owing to the oxidation of the iron. No appearance of this kind is observed with the hydrates of white limes; but on applying a slightly moistened test-paper to the sections, we have at once an evidence of the breadth of the carbonated parts.[d]

[c] "The consistency of a good mortar should be such as to be capable of supporting, without very sensible depression, an iron needle of a line diameter loaded with a quarter of a pound."—*Raucourt sur les Mortiers.*

[d] This operation, however, although simple, is difficult to perform; for the whole sectional surface being hidden by the test-paper, the eye is deprived of the comparison between the operation of those parts which affect it and those which have no action. At the same

83. The annual progress of the carbonic acid is maintained with decreasing rapidity. In fact, the greater the distance the part acted on by the carbonic acid is from the surface, the more difficulties does the regenerating principle meet with in reaching it. These difficulties further vary with accidental circumstances, and with the more or less compact texture of the surface. (App. XXXI.)

84. 2nd. The hardness acquired by the hydrates, varies with the mode of slaking made use of. The three common processes bear the following relation of superiority to one another with regard to rich limes:—1st, Ordinary extinction; 2nd, Spontaneous; 3rd, By immersion. With regard to the hydraulic and eminently hydraulic limes, that order of superiority be-

time, that the difficulty of closely approximating the uneven surface of the cement to the test-paper, more especially in coarse mortars containing gravel, makes the effect, even of the caustic portion, very irregular and undefined; so that it is almost impossible to fix the line of demarcation with tolerable exactness, putting measurement almost out of the question.

The method I have been in the habit of adopting, and which suggested itself to me at the time I first noticed the phenomenon alluded to in the text, will be found more easy in practical application, and satisfactory in its results. It consists in merely detaching a fragment of the mortar to be examined, and immersing it in a neutral metallic solution, containing any dark-coloured oxide, which will be precipitated by caustic lime. The superior affinity of the lime for the acid which is combined with the metallic oxide, immediately decomposes the compound, and the oxide attaches itself to the surface; whereas no action takes place on those parts which have been acted on by the air, the lime being rendered powerless by union with carbonic acid. I found the most convenient test to be a fresh prepared solution of the protosulphate of iron (green vitriol), which deposits the dark

comes, 1st, Ordinary extinction; 2nd, By immersion; 3rd, Spontaneous.

85. On referring to what has been said in Chapter IV., we see without difficulty, that the order of relative hardness is absolutely the same as that of their expansion; that is to say, the method of extinction which reduces the lime to the finest state, is also that which gives the hydrates the greatest strength; a result in conformity with this principle, that the cohesion of a compound ought to be in proportion to the tenuity of its particles, since they can then arrange themselves mutually in the most intimate contact.

86. On consulting the tables in which the experiments are recorded, it is easy to deduce the following conclusions, which are a sequel to the preceding observations:—

1st. Certain very rich limes, free from colour, are capable of forming, by mere union with water, bodies as hard as a number of natural minerals.

green protoxide of iron; but I have also made use of many other solutions and substances with very excellent effect. Some of these are, the bichloride, and the proto, and pernitrates of mercury, producing bright yellow, and brown, or black precipitates, the nitrate of silver, black. Also the infusion and tincture of nut-galls, or of the seed-vessels of the Terminalia chebulica (one of the mirabolans), a plant common in India, which gives a dark green colour (with a strong infusion), in a short time turning to chocolate. With the solution of green vitriol the action is immediate, and when applied to the fresh broken surface of a fragment (not less than half an inch to an inch in thickness, and from six to twelve months old) of a light-coloured plaster, the materials of which have been well incorporated, the separation will be very clearly and beautifully exhibited, the parts acted on by the atmosphere being accurately defined, by a white band surrounding the darker-coloured centre. (Vide App. XXX.)—Tr.

2nd. The carbonic acid diffused through the atmosphere increases, with age, the hardness of those bodies on which it fixes itself.

3rd. The hydraulic limes of all kinds, more especially those which are strongly coloured by iron, cannot, by mere union with water, form any but bodies which are light, and of but moderate hardness.

4th. The carbonic acid augments their hardness also, but never to such a degree as to raise it to an equality with that which it gives to the hydrates of the rich and white limes.

Action of Water on the Hydrates.

87. Water dissolves those parts of the hydrates of rich lime, which are not combined with carbonic acid, whatsoever may be their cohesion.[e]

88. If from the same rich lime, we make, by the three common processes of extinction, three hydrates of the same stiff consistency, and in every respect exactly alike; and if we immerse them immediately in pure water, they will absorb, in a given time proper to each, a certain quantity hereafter specified, viz. :—

89. A thousand parts of the hydrate made by the

[e] When water contains carbonic acid associated with it, which it almost invariably does, it is then capable of dissolving the *carbonate* of lime. This operation takes place extensively in Nature, when the superficial waters of the earth, in filtering through calcareous strata, carry with them carbonate of lime in solution, which is deposited in recesses and caverns, in the form of stalactites. The same effect may be observed underneath the arches of some bridges; though it is probable that, in this case, the lime at first carried down is dissolved in the caustic state.—Tr.

first process, will after one month take forty parts of water, which is the limit of saturation.

90. A thousand parts of the hydrate made by the second process, will after two and a half months take one hundred and eight parts of water, which is the saturating limit.

91. A thousand parts of the hydrate made by the third process, will after two and a half months take two hundred and forty-six parts of water, which is the limit of saturation.

92. The duration of the operation depends upon the mass of hydrate immersed. When we place the materials for experiment in an impermeable vessel, of such a kind that a very small quantity of water suffices to cover it, then the lime dissolved by it, and which it is necessary to take into account, forms only a very minute fraction of the total quantity of the hydrate.

93. It evidently results from the preceding comparisons, that the deficiency in expansion which accompanies the use of the second and third processes of extinction, enables the same lime to be reduced to a paste at first, with a much smaller quantity of water than is necessary to its complete saturation; but we also see that they retain the faculty of completing the dose by an internal and progressive action, yet unaccompanied by any change, except an alteration of density, and by consequence, of hardness; for after the cessation of the action of which we have spoken, no sensible alteration of volume can be perceived.[f]

[f] General Treussart details an experiment which affords a very easy demonstration of this fact:—Two mixtures of a pulpy consist-

94. This observation is of much importance: we shall take care to avail ourselves of it when necessary.

95. We have already, in the first chapter, considered the hydrates of hydraulic limes immersed in the condition of paste, and the hardness they acquire in consequence of this immersion. It remains now for us to point out that this hardness, as well as the quickness of setting, is influenced by the method of extinction, in a manner the more remarkable the less energetic the lime is. The order of pre-eminence of the three modes of extinction is not constant: it is the same for the rich, and slightly hydraulic limes; for the hydraulic, and eminently hydraulic limes, the order is the same as that given in treating of the action of the air.

96. The hydrates of the hydraulic, and eminently hydraulic limes, immersed in the condition of very soft pastes, reject a portion of the superabundant water they contain, and become solid: in the condition of very stiff paste, on the contrary, they absorb an additional quantity, solidify more quickly, and in time acquire a degree of hardness, which lime immersed in the soft state never attains.

97. All hydraulic lime which has hardened in the

ency were made, the one composed of a rich slaked lime, which had been tempered for four years, and the other, of lime fresh slaked as it came from the kiln, both mixed with water. In the course of time, the second mixture became very thick and stiff, the first retaining its fluidity unaltered: water was added to dilute the thickened one to the same consistency as the other; and it again, though more slowly, became agglutinated; and this was repeated several times, before it could retain its fluidity unaltered like the first.—TR.

air, may afterwards be immersed with impunity, without the water dissolving any sensible quantity of it.

98. The hydrates of the rich limes are not of any great benefit in the art; (App. XXXII.) the principal difficulty consists in the shrinkage which the mass undergoes on drying in the air, more especially when it has been slaked by the ordinary process. When we mould it into prisms, of which the dimensions do not exceed those of a very small brick, and if we place them at intervals on an area to which they do not adhere, the mass concentrates itself without obstacle, and the process of drying and solidification proceeds tolerably well; nevertheless the prisms become covered with a slight efflorescence, arising from small portions of the surface which have been unable to participate in the general movement in shrinking. But when the dimensions are enlarged, when the forms become complicated, and the use of moulds becomes necessary, then the material is cramped, the paste adheres to the sides, cracks show themselves, and we get nothing but fragments. Beating is not of any avail whatever.

99. The shrinkage is the more considerable, the richer the nature of the lime employed, and the more it has been swelled in slaking.[g] But, if we could con-

[g] It appears, that when slaked with much water, the lime shrinks on drying, because it is incapable of combining with the fluid which forms it into paste; and its particles must therefore necessarily approach one another when it is withdrawn from between them by evaporation. It shrinks least when made into paste after being slaked by exposure to the air, for then all the interstitial fluid may be solidified by the lime itself, and it is probable that the expansion of the particles of lime during this absorption fills up the spaces

fine ourselves to simple forms, we might succeed in manufacturing at slight cost, with the rich and very white limes, small slabs, which being susceptible of a good finish, and polished on a fine free-stone, would imitate handsome white marble, and might be applied to many uses.

100. The hydraulic mortars in the condition of hydrates, can only be employed with success under ground, or in the water, whether it be in puddling, or in masses of beton. But they will not succeed better than the mortar which is obtained by mixing them with common sand; so that it would be very bad policy to neglect that means of at least doubling the bulk of the material.

left by the abstraction of the water. (Art. 93.) It must not be forgotten, that the remarks contained in this Chapter refer merely to the solids formed of lime and water, *without the admixture of sand* or other ingredient.—Tr.

SECTION II.

CHAPTER VI.

OF THE MATERIALS WHICH ARE ADDED TO LIME, IN THE FORMATION OF MORTARS OR CALCAREOUS CEMENTS.

101. These materials are, 1st, The different kinds of sand, properly so called: 2nd, Arenes: 3rd, Psammites:[a] 4th, Clays: 5th, Volcanic or pseudo-volcanic products: 6th, Artificial products arising from the calcination of the clays, the arenes, and the psammites; and the rubbish and slag of manufactories, forges, glasshouses, &c.

INGREDIENTS OF MORTAR.

Sand.

102. The granitic, schistose, and calcareous rocks, the free-stones ("*grés*"), &c., &c., reduced to the state of hard and palpable grains, either by the agitation of water, or by spontaneous disaggregation, give birth to the various kinds of sand. We distinguish them from powders, by their at once falling to the bottom

[a] These terms are explained in Articles 109, 110, 111, and 112.—Tr.

when thrown into limpid water, and that without altering the transparency in a sensible degree.[b]

103. The disaggregation of rocks is often accompanied by a decomposition which produces a powder. This powder renders the sand "rich," or, in other terms, susceptible of a certain cohesion, when tempered with water; washed by rains and currents of water, it is soon freed from the pulverulent particles, and is deposited pure in the beds of rivers.

104. This purity is generally changed near the mouths of streams, and in the small rivulets whose tributaries flow over a bed of clay or mould; the sand mixes with vegetable debris and animal matters, and becomes "loamy." The particles composing sand faithfully represent those of the rocks whence they are derived. The granitic regions furnish quartz, felspar, and mica; and the volcanic regions, lavas of all kinds. The tabular-shaped sands, whose particles are tender, are furnished by the schistose mountains; it is difficult for them to be transported far without being reduced to powder.

105. The calcareous sands are the least common. This very probably arises from the circumstance, that

[b] The specific gravity of siliceous or quartzose sand I found, by a mean of various experiments, to be 2.6, which is the same as that of quartz. A cubic foot of this, if it were *solid*, would therefore weigh 162½ lbs., since a cubic foot of water weighs 1000 ounces. But I found that, when measured in the way usually adopted in apportioning the ingredients of mortar, (in which case the sand, which was usually rather damp from being fresh dug, was thrown into the measure, but *not beaten or compressed*,) a cubic foot of it weighed only 75 lbs. Hence the *voids* or spaces between the grains were rather more than equal to the bulk of the sand itself, when merely thrown together. The same method may be employed to ascertain

there are but few rivers but what take their rise from primitive summits, or such as are composed of primitive elements. The calcareous rocks, besides, are not susceptible of that kind of disaggregation which could be called granitic; for if they be of a soft kind, they merely produce powder; if hard, scaly splinters.

106. The partial and secondary revolutions of the globe, have occasioned immense deposits of sand in situations where now neither brooks nor rivers flow: these are the fossil sands; and they should be carefully distinguished from the virgin sands, which are still in their original scite, and have not been operated on by the waters.

107. The fossil sands generally exhibit a more angular grain than the sea or river sand; but in other respects they are the same elements, sometimes pure, sometimes coloured by ochres, &c.

108. Among the fossil sands is one very remarkable, the arene. Its peculiar properties entitle it to a distinct place in our category.

the density of sand in other conditions; and it is equally applicable to any *mixture* of gravel or rubble and sand, if the pebbles and grains contained in it do not vary much in specific gravity from that of flint and other siliceous minerals, which is rarely the case. The knowledge derived from this experiment is of considerable importance, in reference to the established principle, that the best proportion of lime or other cementing matter in a mortar, is that which exactly fills up the voids between the particles of sand with which it is mixed, and which is thus obtained. For the difference between the weight of a cubic foot of the materials, *when condensed as much as possible*, and 162½ lbs., will indicate the proportion which remains to be filled up with cementing matter; which is the *least* dose (supposing the whole of it to be active) that will bind them firmly together, but beyond which as little should be added as possible.—Tr.

INGREDIENTS OF CEMENTS.

Arenes.

109. This is a sand, generally quartzose, with very irregular, unequal grains, and mingled with yellow, red, brown, and sometimes white clay, in proportions varying from one to three-fourths of the whole volume.

110. The arene almost always occupies the summits of the rounded and moderately-elevated hills: it sometimes constitutes entire hillocks; frequently it interposes itself in large veins and seams in the clefts of calcareous rocks: it belongs essentially to alluvial soils. (App. XXXIII.)

Psammites.

111. We apply this term to an assemblage of the grains of quartz, schist, felspar, and particles of mica,[c] agglutinated by a variable cement. The varieties of these are very numerous; those which in appearance strongly resemble the free-stones, and siliceous breccias,[d] belong to the class of rocks whose disaggregation furnishes sand properly so called. But the psammites, which are slaty, of a yellow, red, or brown colour, fine grained, unctuous to the touch, producing

[c] "Schist" is the German term for slate: felspar is "a simple mineral, which next to quartz constitutes the chief material of rocks." (Lyell's Geology.) "Mica is a simple mineral, having a shining silvery surface, and capable of being split into very thin elastic leaves or scales." (Ibid.) It is often mistaken for *talc*, which it resembles.—Tr.

[d] An assemblage of angular fragments glued together by any cement is called a "breccia."—Tr.

a clayey paste with water, form a distinct species, and one which merits our attention.

112. These last belong to the primitive schistose formations; they do not, and cannot exist, except "in situ;" they are found in beds or veins, forming part of the schist of which they are merely a decomposition. The Department of Finisterre, in the neighbourhood of Carhoix and Brest, furnishes it in abundance. (App. XXXIV.)

Clays.

113. Clays are earthy substances variously coloured, fine, soft to the touch, which diffuse in water with facility, forming with it a paste, which, when kneaded to a certain consistency, possesses unctuosity and tenacity, and may be drawn out and kneaded in every direction without separating. The clayey paste, when dried, retains its solidity, hardens in the fire, &c.

114. Clays are essentially composed of silica and alumina: these two substances are adulterated by the presence of the oxide of iron, the carbonates of lime and magnesia, sulphuret of iron, and of vegetable combustible matter partly decomposed.

115. The clays are separated into four classes: viz., the refractory, which resist, without melting, the heat of the porcelain furnaces (140° Wedgwood); the fusible clays; the effervescing, or clayey marls; and, lastly, the ochrey clays, coloured red or pure yellow by the oxide of iron.

116. The position of clays is very varied: we find them as veins in primitive formations; in hillocks, on

the confines of the primitive chains; in horizontal beds, or layers, in the secondary formations; in threads, thin veins, or infiltrations, in chinks and hollows of calcareous masses; lastly, in volcanic regions. There, their formation is attributed to the decomposition of the compact lavas, and perhaps also, with some probability, to miry eruptions.

117. We shall confine ourselves to these short observations; as the mineralogical history of the clays cannot be entered upon here. On this subject, the reader may consult the excellent article "Argile" of M. Brogniart. (Dictionnaire d'Histoire Naturelle.)

Natural Pouzzolanas, or Volcanic and Pseudo-Volcanic Products.

118. Pouzzolana, properly so called, is a volcanic matter, pulverulent, of a violet red colour, first dug out of the earth by the Romans near the town of Pouzzol, not far from Vesuvius.

119. The environs of Rome furnish it equally. The naturalist Faujas de St. Fond, has found it in France, on the extinct volcanoes of Vivarais. There are few regions exposed to igneous agency which are destitute of it. But it presents itself under very different physical appearances; sometimes pulverulent, sometimes in coarse grains, often in slag, pumice, tufa,[e]

[e] "Tuff, or tufa, volcanic. An Italian name for a variety of volcanic rock of an earthy texture, seldom very compact, and composed of an agglutination of fragments of scoriæ and loose matter ejected from a volcano."—*Lyell's Geology*, 5th edit., vol. i. p. 461.

&c. Its colour, which is generally brown, passes to yellow, grey, and black.[f]

120. We shall, in the rest of this work, comprehend under the name of pouzzolanas, the pseudo-volcanic products arising from the ignition of coal-pits, such as the tripolis, calcined sandstones, clays, &c.

121. Of these substances, some appear to have been exposed to a very high temperature; others seem to have been only slightly scorched. Many appear to have been re-acted on, and modified anew by the heat, or altered by the effect of a very slow spontaneous decomposition. These changes have determined in the constituent principles a union more or less close, and consequently more or less difficult to overcome by chemical agents.

122. The pouzzolanas are essentially composed of silica and alumina, united with a small quantity of lime, potash, soda, and magnesia. Iron is associated with them mechanically, in the magnetic state.

123. The mineralogical history of the pouzzolanas will be found in the works of Desmarest, Faujas-de-Saint-Fond, and other authors. (App. XXXV.)

Artificial Pouzzolanas.

124. Under this denomination we shall include the clays, arenes, psammites, and schists, properly cal-

[f] The only preparation this material undergoes previous to use, is that of pounding, or grinding and sifting, whereby it is reduced to powder; in which state it is beaten to a proper consistency with a due proportion of lime. A specimen of it analyzed by M. Berthier contained, silica 44.5, alumina 15, lime 8.8, magnesia 4.7, oxides of iron 12, soda 4, potash 1.4, water 9.2, in a hundred parts.—Tr.

cined; smithy slag, the refuse of the combustion of turf and coal, and pounded earthenware; and, lastly, tile and pot shards.[g]

125. Such is a brief account of the substances which unite with lime in forming calcareous cements. But these bodies, although generally composed of silica and alumina, are far from acting in a uniform manner; some bind well with the rich limes, others with the slightly hydraulic, or eminently hydraulic limes; and of these several mixtures, some resist the effects of the air, the weather, and the action of water; others only retain their solidity while immersion continues: and again, there are some which

[g] Mr. Smeaton tried powdered forge-scales, such as fall from iron at a smith's anvil, with excellent effect. Also the siftings of the iron-stone after calcination at the iron-furnaces, called "minion," which were ground in a mill previous to mixture with lime. He considered the forge-scales, when well powdered, and sifted clean from dirt and glassy slag, to be equivalent to as much pouzzolana, or tarras; *rust* of iron, or iron-ore burnt, powdered, and sifted, to be equivalent to *minion;* and each of these last to about half the quantity of pouzzolana, or tarras. In the construction of the new Docks at Sunderland, which I visited during their progress, in May 1835, I found that the cement used there, was a mixture of two parts of lime (which was obtained from a blue species of limestone, probably lias) and one part of the ground slag from the iron-founderies in the neighbourhood. I examined some of the lumps of slag lying about; they seemed to contain much iron, either in the metallic state, or as the protoxide. (Vide note to Art. 138.) Before being applied to use, they are ground on a cast-metal bed to a very fine powder; and I was informed that the ashlar work was mostly all united with a cement of this powdered slag and water, without any admixture of lime. This was kept dry for a couple of days, in which time it set; and it afterwards indurated slowly, becoming of a stony hardness.—Tr.

lose all their cohesion as soon as they are immersed, &c., &c.

126. To what are these differences owing? They can neither be accounted for by the physical character, nor chemical composition of the different varieties; (if we confine ourselves to determining the number and the proportions of the constituent principles;) for in this last respect, most of them will be found to be identical.

CHAPTER VII.

OF THE QUALITIES OF THE DIFFERENT MATERIALS WHICH ARE JOINED WITH LIME IN THE FABRICATION OF MORTARS, OR CALCAREOUS CEMENTS.

127. In what follows, we shall term every substance "*very energetic*," which, when kneaded to a clayey consistency with very rich lime, slaked by the ordinary process, forms a cement or mortar capable, 1st, of *setting* from the first to the third day after immersion; 2nd, of acquiring after one year the hardness of good brick; 3rd, of yielding a dry powder under the spring-saw (a little saw whose blade is a clock-spring).

128. 2nd. We shall call merely "*energetic*" every substance which, under the same circumstances as before, will produce a cement or mortar capable, 1st, of setting from the fourth to the eighth day; 2nd, of acquiring, after a year's immersion, the hardness of a very soft stone; 3rd, of yielding a moist powder with the spring-saw.

129. 3rd. We shall call every substance "*feebly energetic*," which, under the same circumstances as before, produces a cement or mortar which, 1st, will set from the tenth to the twentieth day; 2nd, which acquires, after a year's immersion, the hardness of dry soap; 3rd, which clogs the saw.

130. 4th. Lastly, we shall say that a substance is "*inert*," when its presence in proper proportions in rich lime in paste makes no alteration whatever in the manner in which the lime would behave, if immersed without mixture.

131. In the first three cases, what rigidly characterizes the substance, is the hardness acquired at the period fixed; for the "time of set" may sometimes transgress the prescribed limits. In fact, we know hydraulic cements which set on the third or fourth day, without ever attaining the hardness of soft stone; and others, on the contrary, which become very hard, although sluggish at first.

132. These definitions being fixed, we establish as the result of experiment,

1st. That the sands, properly so called, are generally "inert" substances.

2nd. The arenes, the psammites, and the clays, are generally "feebly energetic," and rarely "energetic" materials.[a]

3rd. The pouzzolanas, natural or artificial, may

[a] I have met with a clay of a common kind which, when powdered in its *crude* state, and made into a stiff paste with rich slaked lime, set under water in ten days; but it soiled the finger on first hardening, was meagre and gritty, and did not possess much cohesion. In five months and ten days, however, it had very much improved, and seemed to promise to turn out an excellent cement. I was, however, unfortunately compelled to abandon it at that time, and have been unable to learn its subsequent progress. Its hardness was *then* indicated by a depression of .2125 inch, when tried by an instrument similar to that used by Mr. Vicat, but no part of the surface was removed previous to subjecting it to the blow of the proving needle.—Tr.

be "very energetic," or "simply energetic," or "feebly energetic."

133. There is nothing in the physical characters, either of the arenes, the psammites, or the clays, which will enable us to prognosticate with entire certainty, what their action on rich lime will be.

134. In this respect, the natural and artificial pouzzolanas offer some indications, but they are merely negative. Thus the hardness of grain, density, vitreous or enamelled aspect, want of adhesion to the tongue, in a word, everything which indicates a great cohesion, is an almost sure sign of mediocrity. With these substances then, as with the different kinds of lime, it is still experience which must guide the builder. Nevertheless those who possess chemical knowledge, may apply it usefully in this case; for without making a rigorously exact measurement of the qualities of the above-mentioned substances, these agents assist us in classing them in an approximate manner, by pointing out, to a certain extent, the state of combination of their principles; it is this state which we shall endeavour briefly to investigate in what follows.

The Action of Acids.

135. The pure sands, which are chiefly quartzose, are not acted upon, either with heat or cold, by the common acids (sulphuric, nitric, and muriatic), even when most highly concentrated: this is not the case when they contain particles of very hard volcanic or pseudo-volcanic mineral; but even then the action of acids is extremely slow, and very partial, (the calca-

reous sands, or those mixed with calcareous matter, being excepted.)

136. The clay of arenes, when separated from the sand by washing, dried, pulverized, and set to macerate for several days in muriatic acid, parts with a portion of its iron, and moreover from one-tenth to three-fifths of its alumina.

137. The schistose psammites are exactly in the same position.

138. Muriatic acid dissolves a large portion of the oxide of iron,[b] and nearly all the carbonate of lime in common clays, when they contain them; but it does not act upon their alumina with anything like the same energy, except in certain arenes and psammites. The quantity dissolved rarely exceeds one-fifth, after two or three days' digestion, and often does not amount to one-tenth of the whole.

139. The action of acids upon the natural pouzzolanas is extremely varied; some of them are quite refractory; others yield up a good quantity of oxide of iron, and more than half of their alumina, even in dilute acids. There are kinds, which when treated with sulphuric acid, become in a short time covered with an aluminous efflorescence, and others exhibit no change after a month, or even much more. (App. XXXVI.)

140. The artificial pouzzolanas may exhibit exactly the same phenomena. This observation sufficiently

[b] "As far as is known at present, iron combines with only two proportions of oxygen, and forms two oxides, the protoxide and the peroxide. The protoxide has a dark blue colour; the peroxide is red."—*Thomson's Chemistry*, vol. i. p. 487.

proves, that the principles of the different substances are in a varied state of combination; that in some, for example, the oxide of iron is merely mechanically interposed, while in others it is more or less fixed by the silica and alumina; that these earths themselves may exist in a state of partial mixture; and lastly, that sometimes the whole of the elements are closely and intimately united, by a strong chemical cohesion.

Action of Lime-water.

141. Lime, even in a boiling state, is destitute of action upon the quartzose or calcareous sands. (App. XXXVII.) Such is not the case with regard to the arenes, the psammites, the clays, and the natural and artificial pouzzolanas. These substances, if thrown in the state of powder into limpid lime-water, decompose it more or less; and when in sufficient quantity, reduce it to a state of purity.[c]

142. The following, then, are the conclusions which result from a comparison of the actual qualities of the ingredients of hydraulic cements, and the behaviour of these ingredients, with regard to acids and lime-water.

1st. All the substances which we have termed

[c] This may perhaps be due to the iron present in these substances; for I found that peroxide of iron, prepared from green vitriol, after being washed till it gave no trace of precipitate, on adding the muriate or nitrate of baryta, entirely deprived lime-water of its causticity, on being shaken up, or boiled with it for a short time. Bone-ash and phosphate of lime also have the same property; but silica, even in the gelatinous state, does not affect it.—Tr.

"inert" (except the calcareous sands), are in general entirely unaffected by acids, and have no action upon lime-water.

2nd. Most of the substances which we have termed "feebly energetic," are slightly acted on by acids, and restore a very small quantity of lime (water) to its pure state.

3rd. Most of those substances which we have called "very energetic," are powerfully acted upon by acids, and can restore a pretty large quantity of lime-water to a state of purity.

143. In using the expression "most of the substances," we would have it sufficiently understood that there are exceptions. Hence we see why we should always wait for proof by experiment, when we wish to class a pouzzolana, an arene, or a clay, with perfect certainty.

CHAPTER VIII.

MANUFACTURE OF ARTIFICIAL POUZZOLANAS.

144. The object which we here aim at, is that of replacing the volcanic, or pseudo-volcanic pouzzolanas, by artificial products of at least equal quality, at a slight expense. (App. XXXVIII.)

145. Amongst the rocks, or earths, essentially composed of silica and alumina, those which have been selected as lending themselves with the most facility to the transformation which we have in view, are, 1st, the clays; 2nd, the brown or yellow schistose psammites, which form a clayey paste with water; 3rd, the arenes rich in clay; and 4th, some species of schists.

146. The agent employed is heat; and the conditions of the change are, 1st, that the material shall acquire so much cohesion, as to be incapable of forming a paste with water;[a] 2nd, that it should reach its minimum specific gravity, and its maximum of the faculty of absorption; 3rd, that it become more subject than before to the influence of chemical agents, such as the weak acids. (App. XXXIX.)

147. We fulfil these conditions by the aid of a very moderate roasting, so managed moreover, as to enable

[a] "When the clay contains more than one-tenth of carbonate of lime, it is apt to be deteriorated in quality by a high temperature, and will form a much better pouzzolana when slightly calcined. When it contains little or none, it is improved by an active heat."—*General Treussart.*

us to keep all the particles exposed to the air in a state of incandescence. (App. XL.)

148. The first and best method, consists in previously pulverizing the clay, the psammite, or the arene, which we have selected, and then strewing a layer of it, about a centimetre (four-tenths of an inch nearly) or more, on a plate of iron[b] heated to a point between a cherry red and forging heat. We leave it there till it be raised to the same degree, for a space of time which varies, for each kind of material, from five to twenty-five minutes. We take care to stir the powder continually with a small rod, in order that the

[b] Having, in the course of some experiments on this subject, met with a species of clay capable of transformation into a highly energetic pouzzolana, I was led, in prosecuting a further investigation, instituted with a view to discover the most economical process of conversion, to try the effect of the powder of some broken pieces of a vessel for containing water, made of porous biscuit manufactured from the same clay; and I was much surprised to find it nearly equally energetic with that which had been fresh calcined. The broken fragments of biscuit had lain amongst a heap of rubbish in a damp situation, exposed to heavy rains and the vicissitudes of the weather for many months, *after* the vessel of which they had been parts had fulfilled its office by holding water for perhaps a twelvemonth previous. Yet this powder, *without a second calcination*, when mixed in stiff paste with half its weight of rich lime, set with great firmness in six hours (the same as the calcined clay). We may therefore conclude, that artificial pouzzolanas of this kind are not materially injured by moisture, a fact which greatly enhances their value; and this observation is in correspondence with the experience of General Treussart, who says (p. 123), "As to artificial tarras, when once it has been prepared it requires no further care; for neither the action of the air nor humidity deprive it of any of its properties;" as well as with the results of other independent experiments by myself. (Vide note to Art. 314.)—Tr.

whole of the particles may be uniformly calcined. The clays or ochrey psammites, of a brown, or orange red, or a full blood red, require a heat of a higher degree, and sustained for a longer time, than the others. Twenty minutes, and an incandescence nearer the forging temperature than a cherry red heat, appear to be the limits proper for them. The experiments moreover are so easy, that we recommend every builder to trust, in that respect, merely to his own experience. The mode of calcination which I have just pointed out, has not yet been applied on a large scale; it presents some difficulties. (App. XLI.)

149. The second method consists in rendering the material exceedingly porous, and permeable to the air, if it be not so already, and then roasting it in the manner of bricks, but in the highest part of a lime-kiln, where the heat is insufficient, not merely to vitrify, but even to give a fusible brick the degree of burning necessary for commerce.

150. We make the material porous, by kneading it with an equal quantity of quartzose sand, after which it is divided into loaves or prisms, which are left to dry and harden properly. This plan is particularly suited to the compact and fusible clays; but the pouzzolana which results from it has the defect of being mixed with sand; an inconvenience which might perhaps be avoided, by substituting for the quartzose sand any combustible substances in a finely divided state, such as sawdust, chopped-straw, wheat-chaff, &c., &c.[c] (App. XLII.)

[c] Vide App. X. and its note.—TR.

151. When it is not our object to attain the highest degree of perfection, we content ourselves with roasting the material which we select in the state in which nature provides it. Nevertheless it is proper to reduce it to a small bulk, less than the size of one's fist for instance, if it be not so already; not forgetting that a simple incandescence is sufficient, and that it is by no means necessary to keep it up a very long time.

152. The limekilns, which answer at the same time for burning bricks, draw a great deal of air, and in that respect are perfectly adapted to the calcination of the artificial pouzzolanas. We might also, with equal success, apply the flame which rises above the high smelting furnaces, where they reduce the iron-ore.

153. All clay, principally composed of silica and alumina, and moreover, fine, soft to the touch, and which contains more or less of the oxide of iron, with little or none of the carbonate of lime, will give a "very energetic" pouzzolana, if it be treated by one of the two first methods described above.[d]

[d] Two varieties tried by myself, and calcined according to the first method (Art. 148), afforded excellent artificial pouzzolanas. One of them, a stiff brown clay, which did not effervesce in nitric acid, and which turned brick-red after calcination, set in three hours with half its weight of rich lime. The other, a white kind of pipe-clay, in which silica predominated, but containing no carbonate of lime, and which changed to a light pink colour after burning, set in six hours when made into a stiff paste with lime in the same proportions. After five and a half months' immersion, no impression whatever could be made upon the first of these cements, by the action of an instrument exactly similar to Mr. Vicat's (described at the end of the volume, and represented in plate II.): upon the

154. The same clay calcined in powder in a close vessel, will also give a very energetic pouzzolana; but that pouzzolana will make the cements slower in setting, and more susceptible of alteration by the immediate contact of a flowing stream; while it will also augment the cohesion of its internal particles, and their adherence to foreign bodies. (App. XLIII.)

155. The same clay calcined in fragments, according to the third process,[e] will furnish a pouzzolana simply "energetic;" but if the heat should be pushed so high as to the full burning of brick, we should have only a "feebly energetic" pouzzolana, or perhaps a substance entirely "inert," if, owing to the presence of any fluxes, it should have commenced vitrification.[f]

156. These observations apply only to the clays, the arenes, and the argillaceous psammites. The schists succeed sometimes, but they generally give only mediocre results, more particularly the slaty species.

157. The matters improperly called cements by builders, are artificial pouzzolanas, obtained by the pulverization of old tiles and the rubbish of brick-yards and potteries. Now, as they cast aside among the refuse whatever is defective, either by excess or

second, a very slight indentation was made by the needle, measuring, as nearly as could be ascertained, 0.0125 of an inch.—Tr.

[e] Vide Art. 151.

[f] Five per cent. of carbonate of lime will produce this effect. But the use of such a clay is not to be prohibited on this account, as it is easy to moderate the heat, and its presence is attended with some advantage. General Treussart recommends a rather meagre clay containing the above proportions of lime, in those cases wherein it is impossible to provide machines to effect pulverization, as they are more easily reduced to powder than the richer clays.—Tr.

deficiency in burning; since moreover they employ, for tiles especially, clays impoverished by sand, and more frequently simply stiff earths, we perceive to how many hazards the qualities of these kinds of pouzzolanas are left.

158. The cinders of coal and turf sometimes form an energetic pouzzolana, but they are sometimes also altogether "inert." This depends upon the internal constitution of these substances.

159. The slag from the forge, and the dross from the large furnaces, ordinarily give but "feebly energetic" pouzzolanas.

160. The remarkable cement known by the name of the "aqua-fortis cement," is nothing more than a combination of argil and potash, resulting from a very feeble calcination of nitre and moistened clay. It is a very energetic pouzzolana, but very dear.[g] (App. XLIV.)

[g] Silex, alumina, and baryta, decompose this salt (the nitrate of potash, or nitre,) in a high temperature, by uniting with its base. (Ure's Dictionary.) The potash is rendered caustic, so that it may perhaps act by favouring the partial vitrification of the elements of the clay. Carbonate of soda or potash would, in such case, produce the same effect. (Vide App. XLV.)—TR.

CHAPTER IX.

OF THE RECIPROCAL SUITABLENESS OF THE VARIOUS KINDS OF LIME, AND THE INGREDIENTS WHICH UNITE WITH IT, IN THE COMPOSITION OF MORTARS AND CEMENTS.

161. Let us imagine that we have at our disposal the four kinds of lime described in the first Chapter; and further, all the ingredients mentioned up to Chapter VI.; and, that being puzzled in our choice, we are at a loss to know what course to take. The following are the rules indicated by a very numerous assemblage of facts, collected during observations of fourteen years' continuance. (App. XLVI.)

First Case.

162. To obtain mortars or cements capable of acquiring a great hardness in the water, or underground, or in situations constantly damp, we must combine:

WITH THE RICH LIMES.	WITH THE SLIGHTLY HYDRAULIC LIMES.	WITH THE HYDRAULIC LIMES.	WITH THE EMINENTLY HYDRAULIC LIMES.
The "very energetic" pouzzolanas, natural or artificial.	The "simply energetic" pouzzolanas, natural or artificial. The "very energetic" pouzzolanas, natural or artificial, tempered by the mixture of a half of sand or other "inert" substance. The "energetic" arenes and psammites.	The "feebly energetic" pouzzolanas, natural or artificial. The "energetic" pouzzolanas, natural or artificial, tempered by a mixture of about one-half of sand. The "feebly energetic" arenes and psammites.	"Inert" materials, such as the quartzose and calcareous sands.[a] Slag, dross, &c.

[a] With an eminently hydraulic lime, found in the neighbourhood of Masulipatam (vide note to Art. 24), I found that the mixture of

Second Case.

163. To obtain mortars or cements capable of acquiring great hardness in the open air, and of resisting rain, heat, and severe frosts, we must combine :

WITH THE RICH LIMES.	WITH THE SLIGHTLY HYDRAULIC LIMES.	WITH THE HYDRAULIC LIMES.	WITH THE EMINENTLY HYDRAULIC LIMES.
No ingredient will effect this.[b]	No ingredient will completely effect this.	Any very pure sands. Quartzose powders. The powders of hard calcareous minerals, or other "inert" matters.	Any very pure sands. Quartzose powders. The powders of hard calcareous minerals, or other "inert" matters.

164. If, for any reasons, we should modify the rules above given, we may still succeed in making *tolerable*, and perhaps *good* mortar ; but to a certainty we shall be farther off the *best*, and that in a greater degree, the more completely the combinations made use of tend to invert the scale, which places the "very energetic" pouzzolanas opposite the very caustic rich limes, and

two parts of a highly energetic artificial pouzzolana produced a much inferior cement to a like mixture of the same pouzzolana with *rich* slaked lime. I did not find the time of *set* to differ much, but the cement containing the hydraulic lime was meagre, and friable, and soiled the finger on touching it, for a day or two after solidification ; that prepared with rich lime, formed a compact, perfectly hard mass, with clean surface, and conchoidal fracture, and so homogeneous in texture, and closely united, as to be superior to many substances which had undergone the action of heat, such as bricks, tiles, &c.—TR.

[b] Having analyzed several very old mortars, with the view of discovering, if possible, to what their superior durability might be attributed, I found that the hardest and most compact almost invariably

the "inert" sands opposite the eminently hydraulic, very *mild* limes. Lastly, we shall have done the greatest mischief possible, when we have united together rich limes and any kind of sands. Such is the language dictated by a comparison of facts. It is unnecessary to add, that when we meet with substances of qualities intermediate between those which constitute the categories given above, we must use medium proportions; but fresh experiments can alone tell, what will be the peculiar result of this or that new combination of principles, which it may suit the fancy of the builder to adopt.

contained a considerable quantity of silica in a minute state of division, and a portion in a gelatinous state, which I presumed to have been in chemical combination with the lime. In some excellent specimens of very old mortar also, magnesia was found to exist in considerable proportion. The limestones, therefore, from which these mortars were prepared, must have contained the silica and magnesia as constituent principles; and it is to be recollected, that it is the presence of these substances which communicates the property of hardening under water. Hence we may infer, that the most durable common mortars are such as have been manufactured from hydraulic limes.—Tr.

SECTION III.

MIXTURES OF THE DIFFERENT LIMES WITH SAND, AND THE OTHER INGREDIENTS OF MORTARS AND CEMENTS.

CHAPTER X.

OF CALCAREOUS MORTARS OR CEMENTS INTENDED FOR IMMERSION.

165. Every calcareous mortar or cement destined for immersion, and mixed beforehand as a matrix, with a certain quantity of stones, fragments, or rubbish, constitutes what is called a *beton*. It is a real piece of masonry of small materials, which is deposited at once in the situation which it is to occupy. In what follows, we have merely to consider the *matrix* of the betons.

Choice of Proportions.

166. The proportions which lead to the greatest induration, are as varied as the number of possible combinations of the various known ingredients; that is to say, in a word, the question cannot be well discussed in any but an imperfect and altogether ap-

F 2

proximative manner, and that in this respect, every builder ought to study the materials at his disposal.[a]

167. Of the known materials, the arenes, the psammites, and the clays, appear to have the least affinity for lime; when reduced to a dry powder, and measured in that state, they take, for a volume represented by unity, as follows:—Of rich lime slaked by the ordinary process in stiff paste, from 0.15 to 0.20; of slightly hydraulic lime, from 0.2 to 0.25; and of hydraulic lime from 0.25 to 0.30.

168. The energetic, and very energetic pouzzolanas, require, in the same circumstances, of rich lime, from 0.30 to 0.50; of slightly hydraulic lime, from 0.40 to 0.60.

169. The quartzose or calcareous sands—of hydraulic, or eminently hydraulic lime, from 0.50 to 0.66.

170. In general terms, it is better to err from a deficiency, than an excess of lime, when making mixtures of rich lime, and any kind of pouzzolanas; and *vice versâ* in the case of hydraulic or eminently hydraulic limes, mixed with quartzose or calcareous sands.[b] (App. XLVIII.)

171. The proportions play a more important part, the weaker are the ingredients which are mixed

[a] The reader will find in the notes (App. XLVII.) an extract from Mr. Smeaton's excellent essay on water-cements, containing the proportions of twenty different compositions employed by him, in the construction of the Eddystone Lighthouse and other works, and which, being prepared from materials in common use in England, can not fail to be valuable.—Tr,

[b] "A considerable excess of lime is proper in common mortars; an excess of cement cannot be injurious to hydraulic mortars, except when employed in plastering."—*Raucourt de Charleville*, p. 14.

together; that is to say, that slight differences in these proportions may, in that case, correspond to very considerable differences in the hardness of the compounds.[c]

172. Furthermore, all that we have just said may be modified according to the purpose to which we propose to apply the mortars or cements; more especially in dealing with pouzzolanas and the rich limes. Are these cements intended to bind materials together? Let there be given a slight excess of lime, without which they will not adhere to stone without much difficulty. Are they to be employed alone merely? Let us keep as close as we can to the exact proportions, in order that their hardness may be the greatest possible.

Choice of the Process of Extinction.

173. The nature of the lime, and of the ingre-

[c] Colonel Raucourt de Charleville considers the best hydraulic *lime* to be composed of equal parts of pure caustic lime, and of such other ingredient, as by its chemical action gives birth to hydraulic properties; such as silica, alumina, magnesia, &c., and which he terms "base hydraulique." None however is to be considered such, unless in a sufficiently minute state of division to exert its full influence, and allowance is to be made for all such gritty and inert particles as the best hydraulic lime is apt to contain, their amount being ascertained by experiment, and a proportionably smaller quantity of sand being used in forming the mortar. The best hydraulic *mortar* he considers to be that prepared from a mixture of clean sand with so much of the above matrix as is at least sufficient to intercede its grains. Many hydraulic cements, particularly those made by mixtures *after* calcination, contain so much inert matter as to be incapable of the addition of sand.—Tr.

dients employed, regulates the choice of the process of slaking.

174. Facts lead to the following general observation, viz. :—

1st. That for all possible kinds of cement, of rich, or slightly hydraulic limes, the order of superiority of the three common kinds of extinction is as follows: spontaneous,—by immersion,—and ordinary.

2nd. That for every possible kind of cement or mortar, of hydraulic and eminently hydraulic lime, ordinary extinction,—by immersion,—and spontaneous.[d]

175. There may be exceptions doubtless, but we have to this day met with none.

176. The differences of induration, which follow the use of this or that process, are very variable; they reach their maximum in the case of the rich limes, when mixed with inert substances, and become almost insensible, when these same limes are allied to the very energetic pouzzolanas. Between these limits, the differences observe a progressive scale, which is regulated by the variable energy of the ingredients. (App. XLIX.)

177. We may slake the hydraulic, or eminently hydraulic limes, either by immersion, or by the ordinary process, without entailing any great difference in the mortars or cements in which these limes are employed; but this is not the case with regard to the spontaneous extinction, the influence of

[d] Mr. Smeaton mentions, that the Aberthaw blue lias, when quenched hot from the kiln, falls to powder as freely as any other kind of lime, and that when mixed as fresh burnt as possible, it was thought to set more strongly in under-water works.—Tr.

which is accompanied with a more injurious consequence, the more eminently hydraulic the lime to which it is applied.[e] (App. L.)

Of the Manipulation or Manufacture.

178. The manufacture includes the slaking of the lime, and its mixture with the ingredients which unite with it in the composition of the mortar or cement.

179. In whatsoever way it may have been slaked, the lime ought to be first brought to the condition of a thoroughly homogeneous paste, and then to be mixed with the ingredients destined for it.[f]

180. This paste ought to be as stiff as possible, whenever it is intended to act the part of a matrix amongst hard and palpable grains, which preserve

[e] Mr. Vicat has omitted to mention what seems to me to be an important caution regarding the choice of the process of extinction, viz., that the extinction by immersion will be attended with similar evil consequences, if the lime be exposed to the influence of a humid atmosphere for any length of time; and this is equally the case with the ordinary process, if the mortar be permitted to remain mixed sufficiently long to allow the injurious action of water upon the lime to take effect. Hydraulic lime should be reduced to the utmost state of division, and used as quickly a possible afterwards. (Vide note to Art. 185, and App. L.)—Tr.

[f] In hydraulic mortar composed of lime and tarras, repeated beatings rather improved those compositions in which lime predominated, more particularly when it was rich lime. "The customary allowance for tarras mortar beating, first and last, is a day's work of a man for every bushel of tarras; that is, for two bushels of lime powder with one bushel of tarras."—*Smeaton, Construction of Eddystone Lighthouse.*

a sensible interval between one another. Such is the case with mortars, or mixtures of lime and sand. (App. LI.)

181. It may have a more or less thin consistency, when, with a pulverulent substance, whose grains are impalpable, and at the same time absorbent, we would form a whole of a homogeneous appearance, in which the eye is unable to discern any one of the constituent elements. This is the case with the calcareous cements, or the mixtures of lime with the pouzzolanas, arenes, clays, or psammites. (App. LII.)

182. *But in every possible case, the resulting mixture, be it mortar, or cement, must exhibit a good clayey consistency, according to the definitions laid down in that respect in Chapter V.*[g]

183. We may at pleasure bring lime which has been slaked by immersion, or spontaneously, to the condition of stiff paste or pulp, when we take it in the pulverulent state; but this is no longer possible when we have to deal with lime slaked by the ordinary process, if it has been drowned at first in too much water. To avoid committing this mistake, we ought to employ, at the moment of slaking by that process, no more than the water rigorously required, that is to say, which is necessary to cause the lime to pass from the solid caustic state to that of a stiff paste. In fact there is always an opportunity to add it, if necessary, at the time of applying it to use.

184. The hydraulic, and eminently hydraulic limes,

[g] Vide App. LXII. and its note.—Tr.

present certain difficulties in this respect. We shall here detail the process to be followed, when the ordinary mode of slaking is applied to them.

185. The quick-lime in lumps is shovelled into an impermeable basin, where it is spread out in beds of equable depth from 20 to 25 centimetres (7.8 to 9.8 inches, Tr.); the water is poured in gradually, and in such a manner that it may spread, and easily penetrate the spaces which the fragments of lime leave between one another. The effervescence is not long in displaying itself: we continue to throw in lime and water alternately; but we must take especial care not to mash the materials, and bring them to a pulpy consistency, according to the bricklayers' custom. But when by accident any of the shovelfulls of lime slake to dryness, we turn the water to it by means of little channels, which we draw lightly through the pasty mass, and from time to time thrust a pointed stick into those parts where we suspect the water has been wanting: if the stick comes out of it covered with an adhesive coating of lime, then the extinction has been proper; if, on the contrary, there escape a floury smoke, it is a proof that the lime has slaked to dryness: we then enlarge the hole, make others beside it, and direct the water upon it.[h]

[h] With eminently hydraulic limes, it may be found highly advantageous after slaking, and the separation of the stony and unburnt parts, to bruise and reduce to powder those minute lumps, of which they are often full, as in some instances this operation may have the effect of developing unexpected hydraulic properties.

Colonel Raucourt de Charleville in this manner discovered the virtues of the Narva lime, the best of the Russian cements, which

186. We ought not in this way to slake more lime than is required for the completion of one or two days' work at most. Two separate basins, or two compartments in the same basin, are indispensable. We begin to fill the one, when the other is nearly emptied. By this means the lime has at least twenty-four hours to sour, and the sluggish lumps become all reduced.

187. The lime which has been slaked as I have just described, is already very stiff next day; it is necessary either to pick it, or at least to cut it with a spade, in order to remove it. It seems in a state that it can no longer be brought to a pasty condition without a further supply of water, but that is a mistake; it is easily rendered ductile by means of the pestle. The beater (" rabot") has no longer power to bind it: but if it is battered perpendicularly with rammers of cast-iron, fixed to handles of wood, it is not long in disgorging a part of the water, which, if we may so say, it had rendered latent; it then forms a paste sufficiently thin to receive the sand. (App. LIII.)

188. The materials which combine in the formation of calcareous cements, are worked up and mixed together with the more facility, the greater the quantity of water with which they have been diluted; this is to such an extent, that the same workmen will take four times as much time to prepare a stiff mortar, as would be required to prepare the same quantity of that degree of softness adopted by masons. If, then, it signified but little, in regard to the promptitude with which

had been entirely unknown till he thought of trying this experiment.—TR.

the mortar set, and its ultimate hardness, whether it were worked stiff or soft, it would on the other hand be of importance, in an economical view, to know the limit of the greatest quantity of water proper to be used, in order to keep as near the mark as possible.

189. However, the most exact and varied experiments show, that every mortar or cement destined for immediate immersion, ought, as we have before said, to be worked to a stiff clayey consistency, or lose one-half to two-thirds, and sometimes four-fifths, of the strength it would have acquired if properly treated.[i]

190. But it is not with the ordinary instruments that we can hope to attain the object alluded to. It is absolutely necessary to substitute the rammers we have spoken of above for the beaters, and batter the mass vertically with force and suddenness. It is not until the mortar or cement has been fully worked, that we introduce the rubble or flints which constitute the beton: this second operation also is effected by the aid of the rammer.

Of the Using or Immersion.

191. All the trouble taken in the fabrication is

[i] By leaving a mortar, which has been mixed to too thin a consistency, exposed to the air, and stirring it from time to time, to change the surfaces of contact, till it has become sufficiently firm, it may regain half of the virtues of which the drowning has deprived it. (Raucourt, p. 99). When, however, hydraulic mortars which had become dry, were beaten up again *and water added to them*, they were found to be deprived of hydraulic properties; and this was more remarkable in the eminently hydraulic limes (p. 108). Pouzzolanas are not injured in this manner, until *after* admixture with the lime. (Vide note to Art. 314.)—Tr.

pretty nearly dead loss, if the immersion of the beton be ill done; in no case ought it to be shovelled in. If the depth of the water be inconsiderable, as, for example, a metre, (39.37 inches, TR.) we should lower it to the bottom, and there deposit it gently. The hopper ought to be generally interdicted. The mass of water which it contains, although stagnant, always soaks the beton which passes through it.

192. The box proposed by Belidor is without doubt the best. We have simplified it, by giving it the form of a truncated quadrangular pyramid, inverted. It is suspended a little above its centre of gravity; on reaching the bottom of the water, it empties itself like a tumbrel, turning over by a tumbler movement. The beton leaves it in pyramidal shape, and settles firmly upon its largest base.

193. The immersion of beton is carried on by successive layers, whose depth should not exceed $0^{m}.40^{c}$. (15.75 inches, TR.) As fast as the layer advances within the coffer-dam, or trench, it drives before it a pulp, or milky fluid, which is continually increasing, and becomes the more abundant the greater the breadth and height of the layers are. This pulp, by reason of its fluidity, makes way for the product of each partial immersion, and finally, if we do not take care, rises from one layer to the next, and in such a manner as to leave behind it a thickness of from 3 to 5 centimetres (1.18 to 1.9 inches, TR.) between the two successive beds: a very serious evil; for being composed, as it is, of lime which has been drowned, this pulp never sets more than imperfectly, and thus ruins the continuity of the mass, of which it moreover favours the settlement.

194. It is easy to obviate this inconvenience in a flowing stream; it is sufficient to contrive little openings in the sides of the coffer-dam, and to multiply them to such an extent, that the water of the enclosure may be renewed without ceasing. We must, however, keep them so narrow, as that the beton be not carried through.[k]

195. In stagnant water, we may establish one or two powerful pumps at the extremity of the enclosure, where each layer in it ends, and pump out the pulpy fluid as fast as it arrives. Brooms are employed with success, when the coffer-dam affords an outlet; but this method supposes that we can give each layer time to set.

196. Furthermore, we reduce the formation of this milky fluid to a mere trifle, when we take great care in the immersion. We ought above all to avoid beating the beton when immersed. It is a perfectly useless operation, and is moreover very injurious. The beton of itself settles as much as we could desire; the beating has no other effect than that of diluting and impoverishing it. The only thing which we may allow ourselves to do, is to spread or weigh down the product of each partial immersion by compression, but without giving it any blow.[l] (App. LIV.)

[k] Mr. Smeaton used a coating of Plaster of Paris to guard the cement of the Eddystone Lighthouse, when there was not time for it to set before being subjected to the violence of the sea.—Tr.

[l] The following ingenious method of immersion, proposed by General Treussart, is well deserving of attention, although he does not speak of it as having been subjected to trial. The space on which the foundation is to rest, is first surrounded by sheeting piles, and cleared

197. A practice very generally admitted in former times, and which naturally sprung from the persuasion, that the most rapid set led to the greatest absolute induration, consisted in making use of hot lime, and in immersing the beton while still warm. Now a beton still warm is necessarily one imperfectly reduced; it is in fact quite impossible, that the heat developed by the quick-lime should be retained during the whole period of time required by a good manipulation. Every manifestation of heat at that period

out to the depth of about six feet, which space is filled up to the level of the bed of the river with beton. While this is still somewhat soft, a second row of sheeting piles parallel to the first, and at the distance of four or five feet, is driven into the beton itself, to the depth of eight inches. The two rows are to be firmly bound and united together by struts, braces, and other means, and the space between them filled in with a puddling of clay. Thus a coffer-dam resting on the beton is formed, so that when it has become indurated, the whole of the water may be pumped out, and the masonry of the piers built up within it. But as the removal of the water from the interior may expose the whole to the danger of being lifted up by the pressure of the fluid, since it then becomes similar to an empty caisson, it may be necessary to guard against this danger, by loading it with weights placed upon trestles supported by the beton, which will keep the whole down till the masonry has been executed. When this has been done, the interior line of piles may be easily removed. The outer row will require more trouble if they have been driven to the depth of six feet or more; and there is also a risk of the beton insinuating itself between them, and opposing a further resistance to their extrication; to guard against which, a stout piece of sheet-iron might be bent round the inside of them before filling in the foundation with the beton, which would prevent its entering the joints. Or the outer row of piles may be sawed off level with the beton, and so left, which seems better.—Tr.

indicates the successive developement of a sluggish lime, and consequently an imperfect extinction and mixture.[m] Now if such a mixture be immersed in a spot where nothing confines it, it swells, becomes diluted, and spreads. Hence we must conclude, that it is not proper to use the lime, until after it has completely cooled, which is a certain sign of its being completely slaked. Moreover, it will rest with the intelligence of the builder, to distinguish when the heat given out by a large quantity of slaked lime in paste, is the result of a real internal action, or the remains of the first effervescence. (App. LV.)

198. We ought not, as Belidor directs, to wait to immerge the beton, till it has been deprived of its ductility by an incipient desiccation; for it parts asunder and crumbles in an incredible manner, and will then merely form a pulp, which necessarily sets very slowly and imperfectly: such a beton does not finally attain more than from one-tenth to three-tenths of the hardness of which it is susceptible, when properly immerged.

199. When, from unexpected hinderances in works, we are obliged to postpone the immersion of a certain quantity of beton already prepared, and it has begun to harden in consequence, there is not the least danger in beating it up afresh, and bringing it to its first consistency by the addition of water; provided nevertheless, that the hardness it has ac-

[m] These remarks cannot be applied to the employment of those hydraulic mixtures, which are used unslaked, and ground previous to mixture, such as our Roman cements, concrete, &c., &c.; the theory of the consolidation in these cases, being quite distinct from that of the hardening of common mortars and betons.—Tr.

quired merely arise from a too rapid desiccation, brought on in a few hours by the heat of the sun, or a scorching wind.

200. When it is possible to make use of mortars and cements in the dry way, in a drained enclosure, the beton is generally supplanted by masonry. We may then double the future resistance of which these mortars or cements are capable, by leaving them to acquire, previous to admitting water into the enclosure, a certain degree of hardness, not incompatible with an obvious humidity; such, in a word, that the white tint, which is characteristic of dryness, shall never make its appearance.

Action of the Water upon the parts of the Mortars and Cements in immediate contact with it.

201. The parts of mortars and cements in immediate contact with the water, after having acquired a certain degree of hardness, at periods varying with every description of lime and ingredients, sometimes finally retrograde so far, as even to lose the consistency they had at the moment of immersion. They also form themselves a kind of envelope, whose thickness is constantly increasing, and tends to reach the centre. If we remove it by an iron knife by scraping the surface down to the sound part, it forms a second, and so on, one after another.

202. These phenomena, which are very remarkable when we make use of the rich limes and feebly energetic pouzzolanas, are imperceptible in every case, where we use the very energetic pouzzolanas, or the

hydraulic and powerfully hydraulic limes combined with the ingredients proper for them. We see of how much importance this observation is to the durability of works exposed to the dashing and erosion of a current or agitated water; but in a tranquil pool the progress of this species of decomposition has a limit. A very thin blackish or whitish crust insensibly forms on the soft envelope of which we have spoken; behind this sort of shield, the formation of which is due to carbonic acid with which every water is more or less impregnated, the soft parts regain their consistency by little and little, and everything seems to indicate that at some period its solidification will be complete. It is certain that after six years it is still not equal to the centre in hardness.[n]

203. On being submitted to examination, the deteriorated parts exhibit much less lime than the others; what is deficient then, has been dissolved and carried off; it was in excess in the compound.

204. Nature, we see, labours to arrive at exact proportions, and to attain them, corrects the errors of the hand which has adjusted the doses. Thus the effects which we have just described, and in the cases alluded to, become the more marked, the further we deviate from these exact proportions. (App. LVI.)

[n] Mr. Smeaton made some experiments to ascertain whether any difference would arise in the strength of hydraulic mortar when mixed with fresh or sea-water, the compositions being immersed in the same water; and he found, that after a trial of two or three months, those made up with the salt-water appeared, if there were any difference, to have the preference.—Tr.

Influence of Time.

205. Some persons are in the habit of concluding upon the future goodness of a cement by the rapidity of its "set;" numerous facts prove that this indication is not always constant. The second process of extinction, for instance, hastens the set of all betons of powerfully hydraulic lime, but does not lead to as great a degree of hardness as the ordinary process. Besides, there are ingredients whose binding qualities are only developed in a slowly progressive manner, and which nevertheless attain a very high degree of solidification. The time of first setting cannot be taken as an exact prognostic of the future hardness, except when we compare together cements or mortars of the same kind;° thus, for example, when a mixture of 200 parts of pouzzolana and 100 of rich lime takes six days to set, while a mixture of the same consistency of 200 parts of the same pouzzolana with 270 parts of the same lime take nineteen days, it is certain that the first will become more hard than the second. And this will also be true, if, while the proportions continue invariable, the cements differ merely by the quantity of water introduced into them, and if, by reason of this difference, the more stiff one sets sooner than the other.

206. The following remarks are deduced from the

° "I have remarked also, that mortars of hydraulic limes which set very quickly did not exhibit great resistance; but those formed of pouzzolanas, which caused common (rich) lime to set speedily, always gave good mortars."—*General Treussart*, p. 125.

facts which have been observed up to the present day:—

1st. An excess of rich, or slightly hydraulic lime in a cement, retards its set; the proportions most favourable to that set are also those which give the greatest hardness.

2nd. The second and third modes of extinction seem generally more adapted to hasten the set than the first.

3rd. The progress of the cements of rich limes and the energetic or very energetic pouzzolanas, continues still sensible during the third year following their immersion.

4th. The progress of the mortars formed of the hydraulic or eminently hydraulic limes and the quartzose and calcareous sands, is no longer appreciable after the second year of immersion.

5th. Time modifies, but does not invert the relations in respect to hardness which are deduced from a comparison of the three processes of slaking. That is to say, the order of pre-eminence observed at the end of the first year, is still the same at the end of the third, and so on afterwards. (App. LVII.)

CHAPTER XI.

OF MORTARS CONSTANTLY EXPOSED TO THE AIR AND WEATHER.

207. We have already said in Chapter IX., that the only mortars capable of standing the vicissitudes of the atmosphere, and of acquiring at the same time a great hardness, were those composed exclusively of the pure quartzose, granitic, or calcareous sands, and of the hydraulic, or powerfully hydraulic limes. If then in what follows we treat of ordinary mortars, or the mixtures of sand and rich limes, it is because we are compelled to do so to complete the history of the phenomena which we have to describe. For it is our most decided opinion, that their use ought for ever to be prohibited, at least in works of any importance.[a]

[a] In those situations in which it is impossible to avoid the use of rich limes, it may be useful to be aware that their bad qualities may be in some degree corrected, by the use of a comparatively small quantity of the coarsest sugar dissolved in the water with which they are worked up. This substance (or "jaghery") is extensively employed in the East, and with admirable effect; for the common mortars made of calcined shells, when well prepared at first, resist the action of the weather for centuries; and I have no doubt that this is in great part to be attributed to the use of sugar, the influence of which on the first solidification of the mortar is very marked. Even in this country it may occasionally be found advantageous to employ the cheapest sugar, or molasses, when works of importance have to be stuccoed with rich lime; for its aid is chiefly confined to the hardening of the outer surface, which is effected by

208. Among the results which flow from a general comparison of the mortars with the hydrates, or solids formed by the combination of water alone with the various known kinds of lime, the most remarkable and most important may be exhibited as follows:—

1st. That the hydrates of lime which in the open air attain the greatest hardness, are those whose mixtures with pure sands, on the contrary, produce the weakest mortars.

2nd. The intervention of pure sand does not tend, as was before believed, to augment the cohesion of which every kind of lime indifferently is susceptible; but it is injurious to rich limes, very serviceable to the hydraulic and eminently hydraulic limes, and is neither beneficial nor injurious to the intermediate kinds.

209. Sands being, as we have already said in Chapter VII., merely inert substances, it would seem that they ought not to differ in quality from one another,

employing a stronger solution when laying on the stucco, and rubbing it with wooden floats afterwards, which is sometimes continued several hours; and as the quantity of fluid required for these purposes is not great, the expense of the sugar, if purchased wholesale, would not be a very serious addition to the cost of the work. The proportions used in India are various, depending upon the judgment of the workmen. I am unable to state precisely what quantities the solutions usually contain, but as well as I recollect, there need not be more than about a pound weight to every eight or ten gallons of water in mixing the stucco, and the same quantity to two or three gallons for laying on and floating afterwards. After the sugar has been dissolved, and the solution prepared for use, a quantity of fresh quicklime is added, and well stirred with it, so that as much may be taken up as possible; by this means a very strong lime-water is prepared (note, Art. 14), which is made use of in the manner above mentioned.—Tr.

in any way except by the form, the size, and the hardness of their grains. The ancient builders wished us to choose the fossil sands, harsh to the touch, in preference to the rounded and polished sands: they regarded the colour also, rejected the yellow, &c., &c. But the whole of their writings on that subject are so vague, the experiments on which they depended are so incomplete, and conducted with so little method, that we can conclude absolutely nothing from them. One thing which we know to be quite certain, and which we ought never to lose sight of, is this—that there is no sand whatever, be it red or yellow, grey or white, with round grains or angular ones, &c., which can, if it be inert, form a good mortar with rich lime. Whilst, on the other hand, all possible kinds of sand, provided they be pure, that their grains be hard, and do not exceed a certain size, give excellent mortars with the hydraulic and eminently hydraulic limes. Nevertheless, we admit that there are differences in the quality of sands, according as their constituent elements may be granitic, calcareous, schistose, or volcanic, &c.; but these differences being in general very small, we shall in what follows merely attend to those which depend upon the size of the grains. (App. LVIII.)

Influence of the Size of the Sand.

210. We shall call coarse sand, that, whose grains, supposing them round, vary from one and a half to three millimetres (.059 to .118 inch) in diameter; fine sand, that of which the dimensions in the same

way are comprised between one and one and a half millimetres (.039 and .059 inch); and powders, the solid substances of the same nature, whose largest particles never reach the fifth of a millimetre (.00787 inch). These definitions being understood, experiments prove, that the quartzose and calcareous sands take, with respect to each species of lime, the following order of superiority, viz.:—

211. For the eminently hydraulic, and simply hydraulic, 1st, fine sand; 2nd, irregular-grained sand, resulting from the mixture of coarse and fine;[b] 3rd, coarse sand.

[b] The object of mixing sands is, to obtain the greatest possible quantity of matter within a given bulk; and this is attained by filling up the interstices between the parts of a coarse rubble by the finer particles of sand, and then the interstices of both these by the cementing matter. A method of estimating the proportion of these interstices or voids in a mass of sand is given in note to Art. 102; and by the plan therein detailed, the amount of condensation obtained by mixing them, is at once shown by the increase of weight which it produces. When, however, the means for making such a trial are not at hand, another excellent process, described by Colonel Raucourt de Charleville, may be resorted to. To estimate the voids in any mass of stones, gravel, or sand, a measure is taken and exactly filled with it, and a similar one filled with water; part of the water is gradually transferred from the second to the first measure (pouring it over the stones or sand), till it is full. It is then easy to judge by the quantity of water used, what proportion the voids bear to the whole bulk of the sand, and which will evidently be the same as the ratio of the quantity of water taken from the second measure (to fill them up) to its whole capacity. Colonel Raucourt de Charleville found, that rubble, consisting of pebbles of about half an inch diameter, required *half a measure* to fill up their voids; gravel, of grains from one to two lines, five-twelfths of do.; common sand, of grains half a line diameter, two-fifths do.; fine sand, one-tenth of a line dia-

212. For the slightly hydraulic limes, 1st, irregular-grained sand, mixed as above; 2nd, fine sand; 3rd, coarse sand.—For rich limes, 1st, coarse sand; 2nd, mixed sand; 3rd, fine sand. (App. LIX.)

213. The greatest difference in the hardness of mortars of the rich limes, which the use of this or that kind of sand is capable of occasioning, rarely amounts to more than a fifth; but it exceeds one-third with the mortars made from the hydraulic or eminently hydraulic limes. That is to say, if we represent the maximum hardness in the two cases by 100, the minimum will be not far from 80 in the first case, and 60 in the second.

meter, one-third do.; very fine sands and earths, two-sevenths do. The same author gives an excellent practical method of determining the proportions in which gravel, stones, and sands ought to be mixed, and which is very similar in principle to the process above described. A measure is first exactly filled with the gravel or stones, and they are shaken down in it so as to condense them into as small a bulk as possible. By reason however of the interstices which still exist between these stones, and which by the experiments just quoted have been shown to be equal to one-half of the whole bulk, (when they are of an average size of half an inch diameter or more,) it is evident that there may still be added a large quantity of sand, without sensibly increasing their volume. A similar measure full of sand is therefore prepared, and part of its contents by degrees taken out and poured amongst the stones, which are well shaken, to cause the sand to penetrate and fill up the voids perfectly, more and more sand being added, as fast as it settles into its place, until it be observed that the bulk of the whole begins to augment sensibly, which is a proof that all the voids are filled up. The quantity of sand remaining in the second measure, will now indicate what proportion the gravel can bear to be mixed with it, without augmentation of volume; and to this mixture a still finer kind may be added if thought proper, and the quantity which the mixture will bear, estimated in the same way.

214. When the sand reaches that degree of fineness which constitutes the powders, its mixture with the hydraulic or eminently hydraulic limes still produces excellent mortars, more especially when these powders are derived from calcareous substances endued with great cohesion, such as the marbles. The quartzose powders, though not so efficient as sand of the same nature, may still be very advantageously employed; but, it cannot be too often repeated, that the intervention of any loamy or argillaceous particles in the above-mentioned powders, robs them of their qualities. (App. LX.)

Choice of Proportions.

215. In this case (as well as with regard to mortars destined for immersion) we cannot, to speak exactly, lay down a general rule, since each kind of lime behaves, with respect to this or that sand, in a manner which is peculiar to it. Nevertheless we are sufficiently advanced to lay down certain limits, and to determine a few points, so to speak, of the line which it is impossible to trace throughout with exactness.[c]

In order that this process may succeed fully, the two kinds of sand mixed ought never to bear a less proportion to one another in the diameter of their particles than ten to one, as unless that be the case, the finer will not penetrate effectually. When therefore this condition is not fulfilled, or when we want to compare the condensation obtained by mixing different proportions of common gravel and sands, whose parts are of various dimensions, it will be best to make the mixtures first, and then proceed with the experiment by ascertaining the comparative weights of the same measure filled with each, as before explained in the note above quoted (to Art. 102.)—Tr.

[c] Mr. Higgins' Patent Stucco, made from the best *stone* lime,

Case of the Rich Limes.

216. 1st, The resistance of mortars made from very rich limes slaked by the ordinary process, increases from 50 up to 240 parts of sand to 100 of lime in stiff paste, and beyond that decreases indefinitely.

217. 2nd, The resistance of the same mortars, when the lime has been slaked by immersion, or spontaneously, increases from 50 to 220 parts of sand to 100 of lime in stiff paste, and then diminishes indefinitely beyond that proportion.

Case of the Simply Hydraulic Limes.

218. 1st, The resistance of mortars of hydraulic lime, slaked by the ordinary process, increases from the proportion zero up to 180 parts of sand to 100 of lime in stiff paste, and then diminishes indefinitely beyond that point.

219. 2nd, The resistance of the same mortars, when the lime has been slaked by immersion, or spontaneously, increases from the proportion zero up to 170 parts of sand to 100 of lime in stiff paste, and then diminishes indefinitely beyond that point.

slaked by immersion, consisted of fifty-six pounds of coarse, and forty-two pounds of fine sand (both washed), with fourteen pounds of lime, and fourteen pounds of sifted bone-ash; the whole wetted and well incorporated together with as small a quantity as possible of lime-water, and applied expeditiously. When a finer texture was required, the coarse sand was omitted, and ninety-eight pounds of fine sand used, and to this, fifteen pounds of lime, and fourteen pounds of bone-ash, were added.—TR.

220. These results are sufficient to establish the fact, that the best proportions are subordinate, not only to the nature of the lime made use of, but also to the mode of extinction to which that lime has been subjected. We shall see hereafter, that there are a vast number of other considerations on which they depend.

Choice of the Mode of Slaking.

221. We have stated, in speaking of cements and mortars when immersed, that the manner of slaking the lime exerts a very remarkable influence on their hardness. This influence is by no means so marked in mortars exposed to the deteriorating action of the air. We may, however, by using this or that process with discernment, sometimes do as much as double the resistance which we should have obtained by the contrary process, a fact which well recompenses the trouble of making a few experiments.

222. An examination of the Tables, in which the facts which we have succeeded in collecting together on this subject are recorded, points out at once the same laws which we have already remarked, in respect to cements and mortars when immersed; that is to say, that the three processes of extinction, arranged in the order of pre-eminence, are, for the hydraulic, or powerfully hydraulic limes, 1st, the ordinary process; 2nd, by immersion; 3rd, spontaneous; and *vice versâ* for the rich, or slightly hydraulic limes. (App. LXI.)

The Manufacture or Manipulation.

223. Many authors assert, that mortars are very

much benefited by being soured a long time, but without precisely fixing anything. To put an end to all uncertainty in this respect, we engaged in a new course of experiments, which were further called for by the insufficiency of those we had published in 1818.

224. Quartzose and calcareous substances acting in general to more advantage in the state of sand than as powders, with every species of lime, the mechanical effect of a laborious trituration kept up beyond the time necessary for the perfection of the mixture, can obviously be only hurtful. But there is another point of view in which the subject ought to be considered; it is that of the atmospheric influence upon the ingredients of mortars, an influence which a long trituration keeps up and assists, by the frequent succession of fresh contacts.

225. Now we have seen, that the hydraulic and eminently hydraulic limes, when exposed to the air, lose in it a part of their qualities, while the rich limes acquire new ones from it; hence it follows, that a mortar of rich lime is the only one which can gain anything by being a long time soured, and this is also proved by our more recent experiments. They show, in fact, that a mixture of 150 parts of sand and 100 parts of this lime, slaked by immersion, and measured in paste, having been kneaded and worked up afresh with additional water every eight days during five months consecutively, acquired after one year, an absolute resistance of $5^{k}.43$ per centimetre square;[d] while, in the case of the ordinary manipulation, the same mixture only

[d] Equal to 77.3 pounds Avoirdupois per English square inch.—TR.

reached $4^k.14$;[e] but, however sensible this difference is, it is by no means commensurate with the labour which it cost.

226. It is therefore only by means of renewing the contacts, and favouring the action of the atmosphere, that a long-continued trituration can become favourable to mortars of rich lime; and this furnishes a complete vindication of the Lyonese method, which consists, as is well known, in preparing large heaps of mortar beforehand, from which they take successively as much as is wanted for the day's consumption, rendering it ductile by adding water.

227. But we see at the same time, that the same method may be ill-judged in a country where the lime may be hydraulic or eminently hydraulic.

228. Everything which has been said of the slaking of lime, and the consistency of the mixtures, in treating of mortars and cements for immersion, applies exactly to mortars exposed to the air. The mortars of hydraulic lime may lose four-tenths of the ultimate hardness of which they are capable, when, instead of that stiff consistency of which we have spoken, we make use of that adopted by the masons.

229. Mortar in every season ought to be prepared as much as possible under cover, whether it be to avoid the rapid desiccation which takes place in summer, or to obviate the still more serious inconvenience in the rainy season. In the latter case, we ought to deviate a little from the principles which we have laid

[e] Equal to 58.93 pounds Avoirdupois per English square inch. —Tr.

down in the preceding pages, and choose the hydraulic lime slaked by immersion, in preference to that produced by the ordinary mode; and this in order to have it in our power at pleasure to absorb the water contained in the wet sand; without this plan, it is impossible to obtain a stiff mortar.

230. In summer, on the contrary, the lime in paste is not always sufficient to moisten the sand, which is sometimes hot. It then becomes indispensable to add water, but gradually, and with the greatest caution. One could hardly believe, without witnessing it, how very small a quantity is sufficient to drown the mixture. (App. LXII.)

Application.

231. It is quite evident, that a very stiff mortar cannot be used with dry and absorbent materials. When we have materials of this kind, they must be watered without ceasing, and kept in a perfect and permanent state of imbibition. The whole secret of good manipulation and right employ, is condensed in the following precept: "*Stiff mortar, and materials soaked.*" Our bricklayers, on the contrary, seem to have taken for their motto, "*Dry bricks, and drowned mortar.*"

232. It is true, that to build in the way here understood, we must change certain habits; as, for example, never after the first supply have to introduce mortar between stones too close to one another, but to lay under each one, in the first instance, a sufficient

quantity to allow those at the side to supply themselves from it when it is battered down in bedding it.

233. The mason's hands will soon be covered with sores, if he do not at the same time take some precautions to guard himself from the action of the lime. Liquid tar remedies this very effectually;[f] it is sufficient to rub the hand with it frequently during the day: the thin coating which remains sticking to the skin acts as an impermeable glove. (App. LXIII.)

Precautions to be taken after Application.

234. In general, all mortars become pulverulent when, after being applied, they are exposed to a rapid desiccation. The influence of such a desiccation becomes the more fatal, the more eminently hydraulic the mortars are. They may then lose four-fifths of the strength which they would have acquired by drying slowly. It is therefore proper to water the masonry when we build during the hot season; and this in such a way, as never to permit the mortar to whiten, and thus part with the water necessary for its solidification.[g] (App. LXIV.)

[f] The necessity for this preventive is obviated in England by the pallet or board (called the "hawk"), used by plasterers for mixing small quantities of stucco as they apply it, and for catching such as may fall; it consists merely of a piece of plank about a foot square, with a handle about six or seven inches in length, fixed perpendicularly underneath the middle of it, by which it is grasped.—Tr.

[g] It was found by Mr. Higgins to be injurious to chalk mortar to be kept too long in a very damp state. The object here intended, is

Influence of Time.

235. *Mortar a hundred years old is still in its infancy.* This saying of the masons is the result of the daily observations which they have the opportunity of making in demolitions. It is seldom in fact that we meet with good mortars of rich lime, except in the foundations or masonry of buildings of four or five hundred years old. What is it that determines such a tardy solidification? It is a thing which it appears difficult to account for; however, there is one thing which it is not at all difficult to apprehend, and that is, that a mortar which does not harden for four or five hundred years, is to us much the same as if it never hardened at all.[h] (App. LXV.)

236. As regards hydraulic, or eminently hydraulic mortars, numerous experiments which we have collected prove, that when exposed to the air in small bulk, they in a very short time (eighteen to twenty months) attain, if not the ultimate degree of hard-

merely to *retard* the desiccation of the mortar. Colonel Raucourt de Charleville recommends straw-mats to be suspended in front of the walls as the best means of effecting this. It is an important fact too, that the mortars composed of the very hydraulic limes used in the "active" or imperfectly-slaked condition, are best adapted to resist the injurious effect of too speedy drying. This observation of the above author corresponds with my own experiments on this subject.—Tr.

[h] "Colonel Raucourt de Charleville is of opinion, that in climates not subject to frost, atmospheric vicissitudes, so far from being injurious to mortars, increase their induration, and that in a greater degree the less hydraulic the lime of which they are composed."—*Essai*, p. 112.

ness of which they are susceptible, yet at least a condition differing so little from it, that we are enabled to predict with certainty what they will ultimately become.

237. Thus the influence of ages may modify, but not overturn, the relations in respect to durability established by our observations.

CHAPTER XII.

OF CALCAREOUS CEMENTS AND MORTARS SUBJECTED TO THE CONSTANT INFLUENCE OF A DAMP SOIL.

238. WITH respect to cements, we may refer to all that has been said concerning the proportions, the mode of slaking, and the manufacture, in the case of a constant immersion; and with regard to mortars, to all that has just been said in reference to the case of exposure to the weather, with the exception of the following modifications.

Size of Sand.

239. When the quartzose or calcareous sands are of that degree of fineness which constitutes powders, their presence is injurious to the hydraulic and eminently hydraulic limes; and this effect is the most remarkable in regard to the calcareous sands derived from the softest kind of stone.

Process of Extinction.

240. The varieties of hardness resulting from a difference in the modes of slaking employed, conform to the order laid down, but in general in a more marked way than in the case of exposure to the air.[a]

[a] Vide Art. 222.

Proportions.

241. These are essentially modified agreeably to the following observations:—

1st. The resistance of mortars made from very rich limes slaked by the ordinary process, continually diminishes, reckoning from 50 to 190 parts of sand and more, to 100 parts of lime in paste.

2nd. The resistance of the same mortars when the lime has been slaked by immersion, remains very nearly the same, from 50 to 130 parts of sand to 100 parts of lime in paste, and diminishes indefinitely beyond that proportion.

3rd. The resistance of the same mortars, when the lime has been slaked spontaneously, remains very nearly the same, from 50 to 200 parts of sand to 100 parts of lime in paste, and then decreases indefinitely.

4th. The resistance of hydraulic mortars, no matter by what process slaked, is augmented by very small differences, from zero up to 90 parts of sand to 100 of lime in paste, and remains constant from that proportion up to 240 parts of sand.

Employ.

242. Cements intended to bind together the parts of underground masonry, are used with the same precautions, and the same care, as the mixtures of lime and sand. They lay hold but feebly of the stone or brick when compounded in exact proportions. We augment their adherence by giving them a slight ex-

cess of lime, but this is always at the expense of their own proper cohesion.[b]

[b] For foundations and works of that kind, the use of *concrete* as a substratum in dangerous soils seems in England to be fast superseding every other method, and it offers so valuable a resource in so many difficult situations, that I cannot omit the opportunity of doing the reader a service, by giving a short account of its nature and preparation. Concrete is a composition of stones or rubble and sand with fresh-burned *stone* lime (ground to powder without slaking), in the proportions of from one-fifth to one-ninth of lime, to one of the mixture of rubble and sand. These ingredients should be well blended together *dry*, and as small a quantity of water added as will bring them to the consistency of mortar, and then, after turning over the materials with the shovel once or twice, thrown as quickly as possible into the foundation from a height of eight or ten feet. It sets very quickly, so that it is desirable that the mixture should be made at, or close to the height from which it is precipitated, and after being expeditiously spread and brought to a level, or puddled, it ought not to be again touched. The best proportions of admixture of the stones and gravel or sand, are such as produce the greatest possible condensation, and depend therefore upon the dimensions of both; and the principles by which these proportions may be determined with accuracy will be found in note to Art. 211, it being remembered that the greater the condensation, or the greater the real bulk of the materials packed into a given space, the *less* the quantity of cementing matter necessary to bind them together. The concrete used near London is composed of Thames ballast, containing about two of stones to one of sand, which proportion Mr. Godwin, the author of a valuable essay on this composition, considers to be the best; and as the size of all the stones used is limited to the bigness of a hen's egg (all beyond that dimension being broken), it will be seen that this practical result is in perfect accordance with the experiments referred to in the note above quoted. The quantity of lime (if *hot* from the kiln, and perfectly well burnt and pulverized) should be such as is sufficient to form good mortar with the *sand* (only), which is the simplest rule to go by, as by it the exact proportions in

every case can be decided according to the nature of the materials, and the energy of the lime used, according to the experience and judgment of the architect who employs them. I have not heard of concrete being employed under water, but when once consolidated, it may be exposed immediately to its action without risk. In setting it expands in the same manner as hydraulic lime (vide App. XLIX, par. 3), which renders it very valuable in some situations. This increase of dimensions amounts to three-eighths of an inch in a foot in height on the first setting of the concrete, and it continues to expand insensibly for a month or two afterwards. It is proper to add, that this expansion follows a previous condensation of about one-fifth in bulk, by which the ballast and lime are found to be contracted after being incorporated together.—Tr.

CHAPTER XIII.

OF THE VICISSITUDES TO WHICH CEMENTS AND MORTARS MAY BE EXPOSED, AND OF THE CONSEQUENCES.

243. SOME mortars and cements which solidify tolerably in the water, lose part or the whole of their cohesion when exposed to a dry and warm air. This happens generally with those whose elements do not suit one another perfectly; for instance, the cements of rich limes and feebly energetic substances, such as the arenes, the psammites, clays, &c.; the mixtures of feebly hydraulic limes with inert materials, &c. The deterioration is more particularly marked in respect to the exterior surface, which the dissolving action of the water has already caused to undergo an incipient alteration; these parts lose all their consistency.

244. Mortars formed from the hydraulic or eminently hydraulic limes, whose solidification takes place in a damp soil, behave equally well in the open air, and in the water.

245. In general, most kinds of cement which have well hardened under a damp soil, stand equally in the water, but they behave variously in the open air; some resist, others are altered by it. We are unable to say to the presence of what principles these differences are to be attributed. (App. LXVI.)

246. The mortars of hydraulic or eminently hy-

draulic limes, and in general most cements which have acquired considerable hardness in the open air, will retain it indefinitely in the water, or under a damp soil.

247. The simple mixtures of rich lime with inert sands, whose hardness is merely due to their drying in the air, become completely decomposed in the water.

248. In general, all cements of rich limes resist frosts but imperfectly; they give way like hard stones by irregular cleavage. This action of the frost is considerably weakened, and may even be entirely avoided, by mixing a certain quantity of pure sand with the powdery ingredients which are employed in their composition.[a] (App. LXVIII.)

249. All the mortars of rich limes, and coarse and

[a] In India, where the very speedy decay of mortars and cements, in some situations, appears to be attributable to frequent alternations of dryness and humidity (vide note to App. LXVIII.), I found the best preventive was a coating of tar, which I tried with a view to check the transition of moisture. Part of the surface of a very damp wall, the plastering of which had been long remarked for its repeated and very early failures, was newly stuccoed with the usual composition (4 of rich lime to 5 sand), and after it had dried thoroughly, a portion of it, containing about 50 superficial feet, was painted with a double coating of common tar. This was renewed by a third coat about nine months after, the heavy rains of the monsoon having intervened; and at the end of eighteen months, at which time I was compelled to quit the station, the whole was perfectly sound in every part. The remaining parts of the stucco which had not been thus defended by the tar, began to decay in little more than a month after first completion, and having been restored throughout after a total dilapidation, were *again* (a *second* time) in complete ruin when the tarred surface received its third coat; that is, in nine months after it was first applied. (Vide App. LXVII.)—Tr.

very pure sand, resist the winters of our climate when they have attained a certain degree of solidification; in the contrary case, they are variously acted upon, and this in relation to the proportion of lime they contain. Experience furnishes the following indications on this subject:—

1st. Every mortar prepared in the month of April, with rich lime slaked by the ordinary process, is attacked the following winter, when it contains less than 220 parts of sand to 100 of lime in paste.

2nd. It is attacked in the same way, when it contains less than 160 parts of sand to 100 of the same lime, in paste obtained by immersion.

3rd. It is attacked in the same way, when it contains less than 240 parts of sand to 100 of the same lime in paste, slaked spontaneously.

250. After two years the danger is past, and after six years the most severe frosts are powerless, except the elements of the sand made use of be themselves liable to injury, and consequently subject to be indefinitely subdivided, as far as pulverization.[b]

251. With regard to hydraulic and eminently hydraulic mortars, six to seven months' age is sufficient to place them beyond its reach, whatsoever may be

[b] Several authors agree in the observation, that stuccoes which have to bear exposure to the action of the air, wet, and frost, ought not in any climate to be laid on in successive coats as in in-door work, as their adherence to one another is feeble, and they have a tendency to separate. The joints of the masonry should be cleared out, the bricks well soaked with water, and a single coat of stucco of eminently hydraulic lime and gravel mixed with sand applied, and afterwards floated.—TR.

the proportion in which they may be compounded; but they resist nevertheless in proportion to the quantity of sand they contain. That is to say, that the "poorest" are attacked the last, if any can be. (App. LXIX.)

252. In the covered parts of buildings, the mortars of rich limes have not to undergo these vicissitudes, but they do not become better on that account. The maximum limit of sand proper for them in that case, is from 55, 125, or 175 parts to 100 of lime in paste, according as it is obtained by the ordinary extinction, by immersion, or spontaneously. On referring to the maximum limits laid down for the case of exposure to the weather in Chapter XI., we shall easily deduce therefrom, that the same atmospheric changes are favourable to mortars loaded with sand, and on the other hand are injurious to those in which the lime predominates.

CHAPTER XIV.

INFLUENCE OF BEATING ON THE RESISTANCE OF MORTARS IN GENERAL.

253. MORTAR, considered as a plastic material, fit for moulding, may be made to take every possible form in moulds or shapes. We are besides able to give it the appearance of stone, by making it with fine colourless sand, or rather with fine calcareous powders derived from hard stones.

254. Mortar contained in a mould may be beaten or rammed in the manner of pisé,[a] and acquires by that means great compactness; but as an increase of resistance does not always result from this, it is as well to study the effects of this ramming, and to distinguish those cases where it may occasion more evil than benefit.

255. In order that any material be beaten with effect, it is necessary that it should possess a certain degree of consistency, which is a mean between complete pulverulence, and that state of ductility which constitutes a firm paste. In fact, we must be aware that no compression is possible, when the material is

[a] Pisé, a mode of building formerly in use, whereby walls were formed by ramming and beating down earth, clay, &c., between upright planks.—TR.

able to escape from under the rammer; and this is well understood by the builders in pisé, who never employ any but earth slightly moistened. Now, we always have it in our power to prepare our mortar in this way, whether immediately, or far better, leaving the mortar after it has been worked in the ordinary manner, to undergo desiccation to a proper extent.

256. The successive approximation of the particles of the compressed material to one another, necessarily determines a foliated structure, which though it may not be perceived is nevertheless real. Analogy would lead us to conclude, that in every possible case, a body thus formed ought to oppose a greater resistance to a tractile force in proportion as its direction forms a smaller angle with the plane of the laminæ; however, experience shows that this in general does not take place. The following is what has been determined in this respect:—

1st. Beating has the effect of augmenting the absolute resistance of mortars of rich limes and pure sand in every case, but in an unequal manner. The greatest resistance assumes a direction perpendicular to the planes of the laminæ when the mortars are buried in a damp soil, immediately after their fabrication. It remains parallel to these same planes when the mortars have been exposed to the atmospheric influence.

2nd. The effect of beating is not constantly useful to mortars of hydraulic or eminently hydraulic limes, and calcareous or quartzose sands or powders, except in the case when these mortars are used under a damp soil. The greatest resistance is then in a direction perpendicular to the planes of these laminæ, as with the mortars of rich limes; but in the air, the superiority of the mortars which have been beaten, over those which have not, is only exhibited in one direction, and that is parallel to the plane of the laminæ.

3rd. Beating becomes injurious in every case, when the hydrates of the hydraulic or eminently hydraulic limes are employed without admixture, and subjected to the influence of a damp soil; and is favourable to it only in the direction parallel to the laminæ when the stuff dries in the air.

257. In engaging in the researches which form the subject of this chapter, our object has been to study the properties of mortars of hydraulic limes considered as plastic substances. The numerous casts which we have moulded, both in bas-relief, and in alto relievo, prove, as we have said, that mortar receives and retains impressions well. These casts have stood the rough weather of several winters without the least accident; their hardness has continually been on the increase, and a kind of varnish, with which time has covered them, gives them actually so strong a resemblance to common stone, that the most practised eye mistakes them for it.

258. There remains but one problem to be solved, this is, to discover a means of hastening the set of a mortar, without injuring its future qualities; and this, in order to avoid being obliged to multiply the moulds indefinitely for the same casting. This last

point seems difficult; Loriot's process is altogether insufficient, and besides produces only mortars of bad quality.[b] (App. LXX.) The natural cements, which harden almost instantly in the air and in the water, when worked up like plaster (of Paris, Tr.), are subject to the inconvenience of being tinged brown. Such as we could fabricate artificially ("*de toutes pieces*") by calcining mixtures of lime and clay free from iron, (Chapter XV.) do not stand the weather.

259. But henceforth, mortar of hydraulic lime may be employed as a plastic substance in a multitude of cases, in which the number of moulds ceases to be a difficulty. Such is the case when we have to prepare artificial stones bearing mouldings, vases, or ornaments of any kind susceptible of formation by the rectilinear or circular movement of a profile ("calibre"). It is evident that it will then answer to set the mould in a trench, and run the profile along the clayey paste, prepared and arranged for that purpose. The economy which such a process would

[b] Colonel Raucourt de Charleville found that a good liquid mortar for grouting was composed of eminently hydraulic lime and fine sand run into the joints immediately after mixing, and while still fluid; it hardens immediately without shrinking, and solidifies all its water. Smeaton formed an excellent grouting of pouzzolana and lime, in the proportions constituting mortar (1 lime to 1 pouzzolana) well beaten and incorporated together, and then tempered with water till sufficiently dilute to pour into the joints of his work. In the course of a month this composition became of a moderate stony hardness under water, so that he was obliged to break the vessel in which it was contained to get at it.—Tr.

introduce into our ornamental constructions is indeed incredible.

260. In support of what we have just advanced, we may quote the honourable testimony of the Society of Encouragement. It has welcomed our endeavours favourably, and has been pleased to signify its approval, by presenting us with a gold medal, at the session of 29th of October, 1833.

CHAPTER XV.

OF THE NATURAL CEMENTS.

261. WHEN the proportion of clay in calcareous minerals exceeds 27 to 30 per cent., it is seldom that they can be converted into lime by calcination; but they then furnish a kind of natural cement, which may be employed in the same manner as Plaster of Paris, by pulverizing it, and kneading it with a certain quantity of water.

262. There are some natural cements which do not set in water for many days, and some which harden in less than a quarter of an hour; these last are the only ones which have been made use of at present. Though very useful in circumstances where a quick solidification is indispensable, they are far from affording, in ordinary cases, the advantages of hydraulic mortars or cements of good quality. In fact, they merely adhere to the stone owing to the roughness of its surface, and the entanglement resulting from it;[a] and, however dexterous or experienced the workman

[a] This statement must, I imagine, be understood to apply only to cements which harden in contact with the bricks *under water*, because the adhesion of such as dry in the open air is well known to be much greater than what would be caused merely by asperities of the surface. It is not uncommon to see from 20 to 30 bricks stuck to one another by Roman Cement, and projecting at right angles from the side of a wall, as a proof of the excellence of the composition; and an instance has recently been mentioned to me, in which 33 bricks were successfully supported in this manner.

may be who makes use of them, he will be unable to connect the different parts of his masonry in one continuous bond by means of it.

263. That which is in England very improperly termed Roman Cement, (App. LXXI.) is nothing more than a natural cement, resulting from a slight calcination of a calcareous mineral, containing about 31 per cent. of ochreous clay, and a few hundredths of carbonate of magnesia and manganese. A very great consumption of this cement takes place in London, but its use will infallibly become restricted, in proportion as the mortars of eminently hydraulic lime shall become better known, and in consequence better appreciated.

264. Very recently, natural cements have been found in Russia, and in France; we may compose them at once, by properly calcining mixtures made in the average proportions of 66 parts of ochreous clay to 100 parts of chalk. It is fair however to admit, that no artificial product yet obtained has been able to match the English cement in point of hardness.

265. We remarked (in Chapter II.) that the pure calcareous substances when imperfectly calcined, became converted into sub-carbonates, possessed of certain properties. These properties are to afford a

Now, if we assume the weight of a brick, and its corresponding joint of cement, to be six pounds, and their thickness, when the bricks were joined one to another in the manner above alluded to, (in which the longest dimension of the brick was placed vertically,) at 2½ inches, then the cohesive force necessary to unite the first brick to the wall, with sufficient firmness to bear the strain occasioned by the weight of the remaining 32 supported by it, must have been nearly 91lbs. per square inch; or equivalent to a direct load of 3640lbs. upon its whole surface of about 40 square inches.—Tr.

powder, which when kneaded with water in the same way as Plaster of Paris, acquires in it at first a consistency more or less firm, but which does not continue its progress at the same rate.

266. The argillaceous limestones, and the artificial mixtures of pure lime and clay, in the proportions requisite to constitute hydraulic lime by the ordinary calcination, become natural or artificial cements when they have been subjected merely to a simple incandescence, kept up for some hours, or even for some minutes. This result, which has often occurred in the course of our first experiments in burning the artificial hydraulic limestones, has been equally observed in Russia, by Col. Raucourt; and M. Lacordaire, Engineer of Roads, has not only just verified it with respect to the different argillaceous limestones of the neighbourhood of Pouilly, but has also made a useful and happy application of it in the works which he is directing at the junction of the Burgundy canal; both in transforming these limestones into natural cements, and in turning to account the large quantity of half-burnt lime which is found in the upper layers of the kilns, when the intensity and duration of the heat is so regulated, as not to exceed the limit proper for the lower strata of the charge.

267. The history of these new cements will not be complete, till authentic and multiplied experiments shall have established, both the manner in which they behave in the open air, and stand frost, and the degree of adherence with which they unite to the building stone. We hope that Mr. Lacordaire will soon solve these important questions.

CHAPTER XVI.

OF THE ANTIQUE MORTARS COMPARED WITH THOSE OF MEDIUM AGE AND MODERN MORTARS.

268. The Egyptian monuments present beyond doubt the most ancient and most remarkable examples which we can quote of the use of lime in building. The mortar which binds the blocks of the Pyramids, and more particularly those of Cheops, is exactly similar to our mortars in Europe.[a] That which we

[a] By the kindness of my friend Dr. Malcolmson, of the Madras Medical Establishment, to whom I am indebted for many important suggestions and valuable assistance throughout the progress of this work, I have been put in possession of some specimens of mortar taken by himself from the joints of the largest of the Pyramids (that of Cheops), and which, interesting as it is from its great antiquity, is the more so from some peculiarities in its external appearance and its chemical constitution, the former of which I shall here notice; as from the account of it contained in the text, I am induced to think that the cement submitted to Mr. Vicat's examination could not have been taken from the same edifice as that which I am about to describe.

The specimens which I examined were of a whitish, or pale cream colour where fractured, and the broken surface exhibited a number of white specks, which, on attentively observing the action of strong nitric acid upon them under a microscope, were found to be insoluble. As the cement was also studded throughout with numerous crystals of the sulphate of lime, I satisfied myself that these specks were occasioned by the fracture of the smaller grains of it,

find between the joints of the decayed buildings at Ombos, at Edfou, in the Island of Phila, and in other places, gives evidence by its colour, of a reddish very fine sand mixed with lime in the ordinary proportions. The use of cements was therefore already known two thousand years before our time; perhaps it would be easy to carry that epoch still farther back, were we to consult the ancient monuments of India, and the Sanscrit books, if they speak of the ancient relations of Egypt with that country; but this would be to attach too much importance to an inquiry, more curious than useful.

269. In confining the use of cements to filling

and not, as I at first supposed, by the unreduced particles of an imperfectly slaked lime. The crystals above alluded to were of various sizes, the largest I saw being half an inch long by a quarter in breadth; and most of them corresponded in every respect with the well-known characteristics of selenite, but other portions were of a more granular earthy structure, being also opaque. In general appearance, with the exception of containing these crystals, the cement was exactly similar to a common mortar of the present day. I also observed, here and there, fragments of the size of a small pea downwards, of an indurated ochreous clay, in some cases of a dark red, or purplish colour, similar to that of an over-burnt brick—in other places yellow; and in one instance I found this clay surrounded by, and imbedded in the crystals of the sulphate of lime, of which it formed a kind of nucleus. Now, as the analysis of this cement proves it to contain no siliceous matter, I have no doubt that the particles of clay which exist in it are fragments which accidentally adhered to the sulphate of lime when it was dug out of the earth, and that the mortar is a composition of rich lime and coarsely-powdered gypsum, which has been used as a substitute for sand, in the proportion of about one of the former (by weight) to five of the latter. Its appearance quite corresponds with this supposition, and for a cement of that kind it possesses considerable tenacity.

the narrow joints of their courses, the Egyptians appear to have foreseen the opposing influence which an always scorching atmosphere exerts on the hardening of calcareous compounds. Time has shown us, how their prudence, or chance, has aided them in this respect: for the Roman works on the banks of the Nile already leave no traces, whilst, after forty centuries, many of the Egyptian temples exhibit themselves to our admiration untouched.

270. Besides, a masonry of small materials would not suit a people, who covered the walls of their public buildings with bas-reliefs, and thus confided

The specific gravity of a lump about the size of a walnut I found to be 1.98.

The particulars of the analysis (by Dr. Malcolmson) of this interesting specimen will be found in Appendix LXXII., one of the results exhibited by which is more particularly noticed at the end of the note to Appendix XXXI. In the mean time, as the possibility of the formation of the crystals *within the cement*, by the mutual exchange of constituents, of sulphate of soda and the nascent carbonate of lime, (the resulting carbonate of soda being supposed to have been subsequently washed out,) has been suggested, it would be interesting to inquire whether, in accordance with my own explanation, gypsum be procurable anywhere within reach of the Pyramids; and if so, whether, as is frequently the case, it be in connexion with an argillaceous deposit? The absence of sand in a locality where such an abundance is now to be met with, is also a remarkable circumstance.

In reference to the above suggestion, I ought not to omit to mention, that the cement was found to contain a considerable proportion (18½ per cent.) of soluble salts; principally the sulphate of lime, but with also a small quantity of the sulphate, and a trace of the muriate of soda. These last were, no doubt, derived from the water, which I am informed is in no place anything like pure in the plains near the Pyramids.—TR.

to sculpture the history of their manners, their arts, their battles, and their conquests. Unburnt bricks cemented with clay sufficed for simple habitations, and beneath a constantly unclouded sky such a method of building was equally safe, as expeditious and economical.

271. It was in the country of the fine arts, in that Greece so fruitful in ingenious inventions, that industry, favoured by the climate, began to vary the application of calcareous cements, and to apply them to a number of uses, of which Egypt did not furnish an example. At the same epoch in which the roof of the Areopagus, built of earth, was exhibited as a curious antique at Athens, the houses of the simple citizens were ornamented with stuccoes, which for whiteness, hardness, and polish, were comparable to the Parian marble, and their terraced roofs defended them from the action of the elements. They constructed with flints, or other hard stones of small size, walls, which were not inferior in hardness to freestone; and the artificial pavements were brought to such perfection, that in a few instants they would absorb all the water with which they washed them: thus the slaves walked over them barefooted, without being incommoded by damp or cold. Such, in the times of Pericles and Plato, was the progress of the arts among the same people, who seven centuries and a half before contented themselves with erecting a monument of clay on the Sigeian promontory to their bravest hero!

272. Italy soon witnessed the spread of Oriental customs through her. Greek artisans also flocked thither from all parts. The Romans, in turn, were

able to instruct themselves from the writings of Anaxagoras, Agatarchus, Metagenes, Phytheus, Theocides, and others. Fussitius published the first book of Architecture which appeared in Rome. After him came Terentius Varro, Publius Septimius, and lastly Vitruvius, who lived under Augustus, whose architect he was. His work, the only one which has reached us, is the more precious as it contains, according to the confession of the writer himself, all that the Greeks knew of the art of building. Pliny the Elder in his natural history, and Palladius (Rut. P. Æmil.) in his treatise " De re Rusticâ," have added nothing to what Vitruvius had said before them. We are even tempted to believe, that in more than one place they confined themselves to copying him.

273. It is therefore Vitruvius whom we ought to consult, when we want to clear up any point of controversy regarding the architecture of the Greeks and Romans. But the monuments erected by these people speak still more plainly than their books; and what remain to us in that way, are sufficient to clear up all the difficulties started on that subject.

274. The Romans, as we have already observed, considered that lime of the finest quality, which was furnished by a marble of the hardest and most pure kind. Hydraulic lime, judging from the silence of Vitruvius, was totally unknown to them, at least as regards its properties. Thus they were unable to do without pouzzolana, when engaged on hydraulic works of great importance, such as a pier or jetty into the sea. They were besides so well aware that common lime and sand only could never set under water, that

after having laid the foundations of the piers of their bridges, by means of coffer-dams, they still kept the dam empty for two months, in order to allow the masonry time to acquire some degree of consistency. "Relinquatur pila," says Vitruvius, "ne minus quam duos menses, ut siccescat."

275. They did not trust the success of any works, except such as did not require great strength, to pounded brick used in the way of pouzzolana. In general, all their mortars which are exposed to the air are alike; we recognise them by the presence of coarse sand mixed with gravel; the lumps of lime in it are sometimes so multiplied, that it is impossible to attribute them to defective manipulation. The extinction by immersion, as applied to a very rich lime, can alone account for it.

276. The Roman hydraulic mortars are very remarkable, and differ from ours essentially: they are composed, with few exceptions, of pure lime mixed in large proportion with the fragments of bricks coarsely pounded; thus they resemble a breccia of which lime is the matrix. Nevertheless, as the brick could not be broken up without leaving a small quantity of rather fine powder, it follows that the lime in such an aggregate is never white; it is on the contrary tinged slightly red or yellow, according to the colour of the brick made use of.

277. This mortar was usually intended to prevent the infiltration of water; they formed the bottom and side lining of cisterns, fish-ponds, aqueducts, &c. They beat it forcibly and for a long time, and

after having smoothed the surface of it with a sandstone, they sometimes laid over it a plaster, or red paint, whose composition is unknown. After the above explanation, it is evident that the Romans employed lime in paste in the way of lining, and that the spongy dry substances which they introduced into it, were for no other object but to hasten the solidification, by exhausting its superabundant water. We can understand, that owing to the size of these bodies, the absorption could only take place slowly, and after application of the mixture, a circumstance which facilitated its being compressed while in its place, to make up for the shrinking of the lime. Besides, the bits of bricks being insulated, and surrounded by the matrix, did not injure its continuity, as the same brick would have done had it been comminuted into fine powder. Thus the desiccation of the plaster took place, and the compactness and impermeability of its texture were maintained.

278. Everything therefore in the composition of this cement had its use, and the general application made of it, shows that it answered its object well, without however offering a great resistance, if we may judge from the numerous specimens which have come to our hands. The lime in them is in fact hardly harder than chalk. However, it is of this same cement that the Italians of our day manufacture those caskets and snuff-boxes which they sell to the curious; but it is remarked that they never make use of any but the outward and superficial parts of the plasters, which are usually incrusted by a deposit of carbonate of lime.

Now it is this deposit which, when perfectly polished, supports the rest, and constitutes the beauty and curiosity of the work. (App. LXXIII.)

279. There is a prejudice already pretty generally spread throughout France, that is, that the Romans possessed a secret for the fabrication of mortars. Some suppose the secret to be lodged in the choice of the materials, and others merely in the way of applying them to use.[b] The very evident consequence of these two opinions is, that the Roman mortars ought everywhere to be equally hard. Now there are six to one in which this is not true, as may be seen in Table, No. 15. It is besides certain, that the ingredients, lime, sand, and brick, always apparent in these mortars, are absolutely the same as those of the country where the monuments exist; and Vitruvius has saved us the trouble of this observation by saying, (lib. i. chap. 5,) "I do not decide what ought to be the materials for walls, because we do not everywhere meet with such as are most desirable; but we must make use of such as we can find," &c.

280. Some have thought to give a triumphant answer to this, by saying, that from the mere fact of their having existed for eighteen centuries and more,

[b] Leoni Baptista Alberti speaks of the use of oil, which he declares to have been much employed by the ancients. He says, (book iii. chap. 16,) "The work will be more secure still, if between the rubbish and the plaster you lay a row of plain tiles cemented with mortar mixed up with oil:" and again, "The ancients made their shell either of baked earth or of stone; and where men's feet were not to tread, they made their tiles sometimes a foot and a half every way, *cemented with mortar mixed up with oil*."—*Appeal to the Public, Liardet,* 1778.

the antique mortars must be much superior to the modern ones, whose unfitness is shown by the deplorable condition of most of our buildings. In order that this conclusion be just, we ought to compare grand monuments with grand monuments, and the puny unsubstantial buildings, with buildings of the same kind. We could then, even with advantage, put in opposition to the antique mortars those of our old ramparts, and generally those of the grand edifices of the middle ages. (App. LXXIV.) As for the fragile walls of our private houses, they would figure perfectly by the side of those which Pliny speaks of when he says (lib. 36,)—"Ruinarum urbis ea maxima causa, quod furto calcis sine ferrumine suo cæmenta compomentur."

281. On taking a general survey of the different categories of mixtures contained in our Tables, it will be seen that the limits of absolute resistance of mortars of lime and sand vary from $18^{kil}.53$ to $0^{kil}.75$[c] per centimetre square. Now the resistances of building-stone, taking the basalt of Auvergne for the last degree on the scale, and for the first, the limestone which is not sufficiently hard to stand polishing, are $77^{kil}.00$ and $20^{kil}.$[d] (The soft stone made use of at Paris gives hardly $10^{kil}.$[e]) By this comparison we see, that we ought to be very guarded against understanding, in a literal sense, what some authors say of the possibility of compounding factitious stones as hard as flints, by means of lime and sand.

[c] Reduced to English measures these give from 263.75 to 10.675 pounds Avoirdupois per square *inch.*—Tr.

[d] 1,096lbs. to 284.7lbs. per square inch.—Tr.

[e] 142.34lbs. per square inch.—Tr.

282. In calculations into which the tenacity of mortar enters as a datum, when we have neglected nothing in regulating the proportions correctly, and the choice of the mode of extinction, we may reckon, one may say,—

In the case of the eminently hydraulic mortars, on an absolutely resistance of	12.00kil.[f]
With common hydraulic mortars	10.00
With hydraulic mortars of medium quality . .	7.00
With rich limes	3.00
The bad mortars which our builders manufacture do not give more than	0.75[g]

These resistances refer to mixtures continually exposed to the weather, and one year old.

283. The best cements and mortars, of the same age, which have been immersed, or buried in a constantly damp foundation, do not give more than 10kil.00.[h]

[f] It resulted from some experiments made by Mr. White (Phil. Mag. No. 64), that Parker's and Mulgrave's cements, and pouzzolana, are in respect to incompressibility equally useful, and that brickwork constructed with them would bear on each superficial foot, before the bricks would crack, about twenty-three tons; that fifty tons would totally crush such brickwork; that Portland stone of the best quality would not split with less than one hundred and seventy-three tons and a half; and that a bedding or joint of pouzzolana mortar was not destructible with that weight.—Tr.

[g] The following are the resistances in pounds Avoirdupois when reduced to correspond with the English square *inch.*

	Pounds.
The eminently hydraulic limes	170.8
Ordinary..............ditto...............	142.34
Hydraulic limes of medium quality...	99.64
Rich limes.....	42.7
Bad mortars	10.67

Tr.

[h] 142.34lbs. per square inch.—Tr.

CHAPTER XVII.

ON THE THEORY OF CALCAREOUS MORTARS AND CEMENTS.

284. THE solidification of mortars has been at all times the subject of controversy; Vitruvius has entered into the subject, and many celebrated chemists in our days have made it the object of their meditations. It would be irksome, as well as useless, to notice in this place all the whimsical and odd theories which have been published upon this subject; there are some amongst them which it would suffice to name to show their absurdity. We shall confine ourselves in what follows to discussing, as briefly as possible, the most remarkable and plausible hypotheses; and we shall suppose that the reader is in possession of the principal facts mentioned in the course of this work.

285. The cause of the phenomena was first attributed to the regeneration of the lime by the slow and successive action of the carbonic acid of the atmosphere. This opinion, supported by Black, Higgins, Achard, and many others, held sway for a long time. But Darcet, in analyzing some mortars procured on the demolition of the Bastile, found in them only one-half of the acid necessary for the saturation of the lime; and very recently, M. John, of Berlin, has discovered that some very ancient and hard mortars were

far from containing that proportion. After these facts, and our own remarks upon the difficulty which the carbonic acid meets with in penetrating into the depths of masonry, the received explanation could be no longer maintained.[1]

286. The experiments of Guyton Morveau, on the mutual re-actions, in the humid way, of lime-water and the solutions of silica and alumina, in potash or soda, have caused it to be presumed with some probability, that chemical affinity may take a part in good mortars, and that a portion of the alumina and silica of the sand being acted upon by the lime, enters into combination with it. This opinion, which we adopted in our early researches, was also that of Mr. John; but that able chemist was not long in discovering its insufficiency, by assuring himself by direct experiments, that caustic, and even boiling lime, has no action upon quartz. Soon afterwards, upon some objections by Mr. Berthier, we became convinced on own part, by disaggregating some mortars eighteen

[1] There can be no doubt, nevertheless, that common mortars which do not *set*, are very much improved in hardness by re-union with carbonic acid, a fact which is easily proved by comparing the resistance of the interior of a large mass of mortar, of chalk lime, a twelvemonth old, with that of its superficial crust; the former will be found in a crumbling tender state, while the particles constituting the envelope are firmly knit together and compact in texture. In mortars, however, which set, the induration appears to be due to other causes, and they ought, therefore, to be classed in a category distinct from the simple mixtures of the hydrate of lime and inert materials, and their phenomena studied independently, lest the facts which they exhibit should interfere with our attempts at generalization, by their apparent contradiction to truths already established.—TR.

months old by muriatic acid, (the sand of which had been very exactly weighed,) that hydraulic lime is equally without action on the granitic sands.

287. In this state of things the most precise notions we as yet have, of the induration of the mixtures, *mortars*, are confined to the knowledge that there is neither any combination between the lime and the sand, nor any integral transformation of this lime into its carbonate, by a principle derived from without.

288. It remains, therefore, for us to inquire into the possible influence of a mechanical agency of the particles, either considered as the result of a mere interlacement, or as the proper cohesion of the lime, in comparison with its adherence to the quartzose, or calcareous substances embodied by it.

289. The hypothesis of a mere interlacement will not bear examination, since any two bodies whatever when joined by tenons and mortices, and without glue, always separate by an even section, at the very joint, and nowhere else, when they are subjected to a tractive force directed in any way in reference to the joint; whence it follows, that no mortar can be superior in resistance to its matrix.

290. Macquer seems to be the first who endeavoured to explain the resistance of mortar by the adherence of the lime, as compared with its own proper cohesion. "The minute fineness of this substance," says he, "and its extreme division, which reduces it altogether to surfaces, gives it the faculty of applying itself most intimately upon the surface of the sand, and of adhering to it with a force proportioned to the nicety and

closeness of the contact." This able chemist, elsewhere accounts for the superiority of the aggregate over its matrix, by the property which the particles of slaked lime possess, of adhering to hard bodies more exactly than to one another, or, in other terms, by the superiority of its adherence over its cohesion. This system has been developed very recently, though not published, by Mr. Girard, Engineer of Roads and Bridges.[k]

[k] It is surprising that in the construction of a theory based upon these properties of adherence and cohesion, no pains should have been taken to throw any light upon the nature of the properties themselves. With the causes of either the one or the other we are at present quite unacquainted; for though the study of the facts connected with their developement in certain circumstances may sometimes afford a clew to the explanation of individual cases, much must be added to our present knowledge of the subject, to elucidate satisfactorily even the simplest phenomena. From some experiments which I made with a view to investigate the cause of the simple cohesion of pastes (which were *not* capable of absorbing the fluid forming them), and their hardening when dry, I found that the hydrates of lime and alumina, fresh precipitated oxalate and phosphate of lime (containing water), silica when gelatinous, the hydrated oxides, starch, gums, &c., all possessed considerable tenacity when powdered and kneaded with water; whilst the *anhydrous* substances, on the contrary, such as sulphur, carbon, silica, and phosphate of lime freed from water, carbonate of lime, and sulphate of lead, were very feebly coherent. In hardening on desiccation, an evident superiority was exhibited by those substances which possessed the greatest affinity for the fluid; while, on the other hand, substances capable of acquiring considerable hardness with one fluid, were altogether void of it with another. Thus a paste of slaked lime made with water, when dried slowly acquired a certain cohesion, but a similar paste of slaked (rich) lime with *alcohol* fell to powder. The same substance kneaded with a solution of sugar, for which it has a great attraction, acquired a remark-

291. Loriot and Lafaye viewed nothing but the interlacement in the aggregation of mortars; hence all their pains were directed towards the physical perfection of the matrix, to which they strove to give great compactness; whether by the adoption of processes of extinction, which, by partially dividing the substance, caused it to take but a small quantity of water, or by introducing quick-lime in powder after it was made up; experience has not completely justified these attempts.

292. Before going farther, and to avoid complicating this discussion too much, we shall proceed to define adherence such as we should understand it, and to examine Macquer's hypothesis in all its bearings.

293. Adherence is the result of internal and unknown forces which only act in contact; it therefore increases with the polish of surface, when we apply hard bodies to hard bodies, and with the roughness of surface, when we apply a soft or fluid body to a hard one. But the asperities of a surface, although they may increase the number of contacts, can add nothing to the strength of an aggregate, if the cohesion of its gangue be much less than its adherence; for the rupture or separation will constantly take place in the gangue, and all the excess of adherence beyond the limit which marks the inequality referred to, will evidently be superfluous.

294. The first consequence deducible from this observation is, that the size of grain being the same,

able consistency, but with spirits of terpentine, for which it has none, it attained hardly any at all.—TR.

the more or less perfect polish of the grains, in sand of the same nature, ought not to influence the resistance of mortars of hydraulic limes.

295. Aggregates present four extremely remarkable cases:—1st. The matrix possesses the faculty of hardening without perceptible shrinkage, and its adherence is much stronger than its cohesion; let us call this G′.

2nd. The matrix can not harden without shrinking, and its adherence is much stronger than its cohesion; let us call this G″.

3rd. The matrix is capable of hardening without perceptibly shrinking, but its adherence is much less than its cohesion; let us call this G‴.

4th. The matrix will not harden without shrinking, and its cohesion is much stronger than its adherence; let us call this G⁗.

Probable Theoretical Consequences of the First Case.

296. 1st. The matrix G′ must, it seems, always exhibit a less absolute resistance than that of its aggregates, because its rupture always takes place freely in the direction of a plane surface of least resistance, while the aggregate can only be fractured along an irregular jagged surface of larger developement.

297. 2nd. Proportions being the same, the strength of the aggregates ought to be independent of the size of the sand, if the grains are likewise somewhat similar, because the section of fracture will in every case keep the same developement, and since the notches, whether consisting of plane or curved faces, will expose the same number of elements, and under the same inclination.

298. 3rd. In every possible case, an inequality in the proportions of sand should carry with it that of the resistances, inasmuch as the developement of the section of fracture is dependent upon this inequality.

299. 4th. The nature of the sand ought to be perfectly indifferent, *cæteris paribus*, when their proper cohesion surpasses that of the gangue.

300. The hydraulic and eminently hydraulic limes afford the only gangues that we can with entire certainty assimilate to G′. Hence the mortars which result from the use of these limes, ought to satisfy the four consequences above established. Now experience shows, that the second and fourth are constantly and completely invalidated, and that the third is so, in the case of mortars buried underground.

301. We see, moreover, that we can gain nothing by modifying the foregoing consequences; for if, for instance, we reverse the two first, we shall place ourselves in continual opposition to the third and fourth, which are indisputable. It is in fact, because matter cannot lend itself to the mathematical conceptions from which the first three consequences are derived; for in reality, the asperities and incidental irregularities of surface of rupture of a solid mass of the hydrate of hydraulic lime, are very nearly of the same order as those which we observe on the fracture of mortars of sand of the ordinary size. Thus the mathematical distinctions drawn between the developements of these fractures are, physically speaking, null, and consequently in no relation to the enormous differences which experience points out between the effective resistance of mortars and that of their gangues.

Probable Theoretical Consequences of the Second Case.

302. 1st. The gangue G″ ought to form aggregates of less resistance than itself; for the sand being unable to follow the general movement of such a gangue in shrinking, this movement is obliged to be subdivided, so to speak, into an infinite number of partial contractions, whence arises pulverulence.

303. This consideration perfectly accounts for the bad qualities of mortars of rich lime slaked with much water by the ordinary process, and there is no need to search for any other cause.

Probable Theoretical Consequences of the Third Case.

304. 1st. The matrix G‴ must necessarily give birth to aggregates of less resistance than itself, because the interposed grains, from their want of adherence, interrupt the continuity of strength throughout the mass.

305. This consequence is completely justified in the instance of the bastard mortar, or mixture of plaster and sand; as it would also be in the instance of the mortars of rich limes, called Loriot's and Lafaye's, if it were proved that in these mortars the cohesion of the gangue surpassed its adherence. Now this is a circumstance at least doubtful.

Probable Theoretical Consequences of the Fourth Case.

306. 1st. The gangue G⁗ must produce the worst

of all the aggregates, and this evident conclusion is justified by the instance of the mixtures of clay with sand of all sizes.

307. Thus the mortars of hydraulic, and eminently hydraulic limes, are the only ones which cannot be explained by the theory of aggregates. We are therefore compelled to have recourse to other considerations; and most of the difficulties will at once disappear, if we consent to admit, 1st, that the action of adherence is not confined to a superficial effect or to contact, but that to a certain extent it augments the cohesion peculiar to the matrix; 2nd, that the limits of this extent are the more enlarged, the more favourable are the circumstances in which the mixture is placed, to a continuance of the molecular movement which takes place in the gangue; 3rd, that lastly, the increase of cohesion in this matrix is inversely proportional to the distance of its particles from the aggregated body which acts the part of the nucleus.[1]

[1] That some cases of the solidification of hydraulic mortars are to be explained upon the principle here laid down cannot be denied, as this volume contains many facts in support of it, (vide Arts. 309 and 314, with the notes, especially the latter,) which render it impossible to exclude chemical agency from an important place amongst the causes of the hardening of mortars. Neither, on the other hand, can it be denied, that many instances may be adduced, in which the chemical affinity of the component parts is insufficient to afford a satisfactory explanation of the phenomena, and the energy of such cements must therefore be attributed to some other cause. Thus, the setting of Plaster of Paris, and calcined magnesia in powder, in which there is *no* chemical re-action between the solid particles, seems to be more probably due to the solidification of the interstitial fluid (by which they are kneaded into a paste), by its combination with the solid substance of the paste; which combination I

308. We shall proceed to support this hypothesis by facts and considerations of great weight. When we examine most of the calcareous incrustations attached to the sides of caverns, and above all ancient aqueducts, we remark that the density of the layers decreases in proportion as they recede from the part incrusted. This fact is, so to speak, inscribed on many specimens taken from the bed of the aqueduct of Gard, specimens which we have before our eyes. Molecular and successive motion within solid bodies is attested by a multitude of observations.[m] Mr. Arago has given us some unanswerable instances of it in the change of elasticity of steel springs. Why refuse to admit it in gangues which appear possessed of great powers of crystallization, such as the hydrate of hydraulic lime? Is it contrary to the principles of science to suppose, that the film of lime which attaches itself in adhesion to the face of a hard body, becomes itself a hard substance in respect to the film next to it, and that one after another these films may finally adhere one to another by an attraction superadded to the cohesion natural to them? Doubtless no; such an action may even be kept up for a very long continuance, especially when the moist condition of the matrix favours it.

have ascertained to have actually taken place in these cases. The same may perhaps be a correct explanation of the *first* set of Parker's and other cements, of the *concrete* now so well known, and of some of the hydraulic limes (vide note to App. L.), though there are no doubt many of these which combine both causes of solidification. (Vide App. LXXV.)—Tr.

[m] The crystallization of the metallic mirrors used in Telescopes, induced, *after solidification*, by certain processes, is a remarkable example of this.—Vide *London Encyclopedia*, art. Speculum.—Tr.

309. The very remarkable experiments of Mr. Petot, Engineer of Roads, on the relations which exist between the solubilities of hydraulic lime joined with sand, and the proportions of the mixtures, leave no doubt of the influence exerted by the presence of the quartz upon the cohesion of the lime.[n] The conclusions to which that observation leads, are necessarily analogous to those which have been given above.

310. We shall leave it to the reader's care to apply these principles to the various cases of resistance offered by mortars of hydraulic lime; and in respect to mortars of rich lime slaked by immersion, or spontaneously, we shall confine ourselves to remarking, that in lieu of endeavouring to account for their inferiority by the purely gratuitous hypothesis of the

[n] This observation seems to be confirmed by the result of some of my own experiments, although they were undertaken with quite a different view. In endeavouring to account for the absence of carbonic acid in mortars of great age (note to App. XXVIII.), and in searching for the neutralizing power which had supplied its place in part, in depriving the lime of its caustic quality, I was naturally led to inquire into the influence which the presence of the other ingredients of the mortars might exercise over the exhibition of the phenomenon, in the hopes of tracing it to its true cause. In some of the cements water was found to be a component principle, but this was not always the case, others being met with in which the analyses were complete, although there was little or no water. The only remaining ingredient therefore to which the effect could be referred was the silica; and although, in respect to it, I was unable to obtain any direct proofs on which to found the assertion, that silica exercises a neutralizing power with lime in the humid way, yet its affinity for it and the alkalies is so strongly shown by many well-known facts, more especially in its action in hydraulic cements, and there are so many indirect circumstances connected with its

superiority of the cohesion over the adherence of the matrix, it is much better to believe with Mr. Petot, that being deprived of all power of crystallization, the matrix has no occasion for the presence of the sand in aid of its own proper cohesion.

311. Calcareous cements, as we have already remarked, cannot be assimilated to aggregates; their solidification presents phenomena of another order, which have only been viewed in their proper light within the last few years. Rich lime, when lodged amongst the grains of quartz of an ordinary mortar, retains in it its characteristic properties, which are, to be soluble to the last particle in water, and to remain soft for a great number of years when excluded from the contact of the air. When mixed however in certain proportions with an energetic

influence upon these bodies which tend to strengthen the above opinion, that I cannot avoid looking upon the subject as one well worthy of investigation, under which impression I venture to subjoin the following remarks:—In revising the simplest cases of the phenomenon, the facts are as follows,—1st, That a certain weight of lime which is associated with silica and carbonic acid, is found to be perfectly neutral to the action of tests, though there is only sufficient carbonic acid present to combine with *part* of it.—2nd, That silica has a strong affinity for lime in circumstances which may frequently occur in cements; these are, when it is in a minute state of division, especially if gelatinous; a fact which is shown by its communicating to it the property of hardening under water, (App. XXXVII.) under circumstances in which the induration cannot be occasioned by absorption of the fluid, as in the setting of Plaster of Paris and some other cements (note to Art. 307).—3rd, It is remarkable, that many very old mortars, and particularly those which considering their age present a striking deficiency in carbonic acid, are very difficult of solution in the cold, even in strong acid, some even requiring days

pouzzolana in a pulverulent state, the lime disappears in a way, becomes insoluble, and communicates to the compound the faculty of hardening in a short time, either in water, or in enclosures impermeable to the air. Now, in what way does this lime thus change its nature? The ancients appear to have attributed a part of these phenomena to the faculty of pouzzolanas to absorb a large quantity of water; but it is evident that that effect is null when the absorption of the pouzzolana is complete. Now mix a rich lime in paste with a pouzzolana soaked to saturation, and immerse the mixture; it will harden none the less after a few days.

312. On attentively reading over Mr. John's observations upon the efficacy of pouzzolanas, which he compares neither more nor less to every kind of sand,

before yielding, and parts being occasionally quite insoluble except at an elevated temperature.—4th, What has contributed still more to turn my attention towards the possibility of a neutralizing agency in the silica, is the remarkable fact exhibited by a comparison of mortars of the same age, and of similar compositions of lime and sand, but of different proportions, and which show a manifest relation between the quantity of silica, and the quantity of carbonic acid, the latter being always in the smallest proportion, when the former was in excess, and vice versâ. I have extracted from my notes the following analyses in illustration of this, and they are the only ones strictly comparable with reference to the point in question, as in them the cements which are ranked in the same series and opposed to one another, are all taken from the same part of the *same* building, and are consequently of the same age, and have been subjected to the same exposure. I should also mention in referring to them, that the analyses were all completed without reference to the point now under examination; as it was not till long afterwards, that a comparison of the results led me to observe the relation I have stated.

and more especially this singular assertion, "That if the pulp of lime cannot harden of itself, and without any addition, neither will it be able to do so a bit the better by any admixtures," we saw at once that that scientific chemist had not studied water-cements, and thus that on this head his authority could be of no service.

313. Mr. Berthier, Engineer in Chief of Mines, who was perfectly sensible of the strangeness of Mr. John's

Table extracted from the analyses of various cements (Table 17), showing the relative proportions of lime, sand, and carbonic acid contained by them:—

Series.	No. of Analyses in Table 17.	Age. Years.	Lime	Sand.	Proportion *per cent.* borne by the carbonic acid contained in the cement, to the full saturating dose.
1st. Mortar of Rich Lime.	1	120	1	1.9	92.0
	2	do.	1	2.1	93.8
	3	do.	1	5.9	83.3
2nd. Mortar of Rich Lime.	4	200	1	1.6	93.4
	5	do.	1	1.7	93.2
	6	do.	1	1.8	91.2
	7	do.	1	1.9	91.05
	8	do.	1	3.3	77.4
3rd. Mortar of Rich Lime.	9	150	1	1.83	79.0
	10	do.	1	2.93	76.9
4th. Mortar of Magnesian Lime, consisting of 2 Lime to 1 Magnesia.	11	400	1	2.2	87.0 } The carbonic acid in these two cases amounts merely to the per centage named of the saturating dose for the lime *only*, leaving the magnesia entirely destitute of it.
	12	do.	1	3.7	72.7

The quantities of sand in the above Table have been reduced to the unit of lime to render comparison more easy; by which means

opinion, has endeavoured to explain the solidification of hydraulic cements, by the intervention of carbonic acid, held in condensation in the same way as other gases, in the pores of pouzzolanas and analogous substances: but analysis does not confirm this supposition; on the contrary, it proves that there is very little carbonic acid in the greater part of water-cements. For instance, Mr. John found only 2.25 of acid to 32.76 of lime contained in 100 parts of tarras cements. Besides, how could we explain the deterioration which, with certain water-cements, follows a solidification already far gone? By virtue of what affinity does the lime, when once carbonized, get rid of its acid in order to regain its solubility?

314. We persist in thinking, as we have always maintained till now, that the lime in cements of natural or artificial pouzzolanas, as well as in cements formed with the uncalcined psammites and arenes, enters into chemical combination with these substances.[o] Our

the remarkable correspondence between the proportions of silica and carbonic acid are rendered obvious. Indeed, the surprising accordance to the supposed relation, which is exhibited by the first four numbers of the second series, is I have no doubt accidental, as the experiments were not made with the scrupulous exactness necessary to entitle them to confidence in such minute particulars; but I cannot attribute to that cause the close correspondence to it exhibited by the sudden change in the proportion of both substances in the fifth analysis of that series.—TR.

[o] A remarkable confirmation of this opinion was exhibited by the result of some experiments on artificial pouzzolanas made by myself. From these it appeared, that no absorption of fluid whatever took place during the set of a cement composed of one part of well-tempered rich lime to two of an excellent artificial pouzzolana. I also found that this powder, which when used fresh

opinion harmonizes with numerous facts set forth in the course of this work; these facts, it is true, cannot be looked upon as direct proofs, but we know, that in matters of this kind direct proofs are extremely difficult, and sometimes impossible to obtain; we know, moreover, that geometricians consider two straight lines to be equal, when they have proved that one can neither be smaller nor larger than the other. (App. LXXVII.)

with half its weight of lime set firmly in six hours, was not perceptibly impaired in energy by an immersion in water of a month and eight days; but a similar artificial pouzzolana, immersed *after* being combined with lime, and formed into cement, was very nearly deprived of its virtues in little more than the same period, when tried by being a second time reduced to the finest powder, kneaded into a stiff paste, and put under water. (App. LXXVI.) The same opinion is entertained by General Treussart; for he says (p. 143), "In accounting for the solidification of mortars under water, it seems to me that we ought to divide them into two classes, altogether distinct; those which are composed of hydraulic lime and sand, and those composed of rich lime and pouzzolana." And again, (p. 144,) "As to mortars of rich lime and pouzzolana, I do not see how we can explain their solidification under water, without admitting a combination between the lime and the pouzzolana." And I have been led to the same conclusion myself, though adopting a different explanation of the hardening of hydraulic limes from that of the author just quoted. (Vide note to Art. 307, and App. LX.)—Tr.

APPENDIX.

NOTES ON CHAPTER FIRST.

(I.) We find the history of lime in all the systems of chemistry; considered as an elementary substance till 1807, it now belongs to the category of metallic oxides, under the name of the oxide of calcium. When associated with carbonic acid in the proportion of 127.4 to 100, it constitutes the carbonate of lime of mineralogists; it is a substance the most widely diffused in nature, and without doubt the most varied in its form, texture, appearance, and association. Carbonate of lime, in fact, belongs to every formation and every epoch of the globe; it is associated with the primitive rocks, forms a part in the transition regions, and composes a large portion of the secondary and tertiary formations. The coral worms create this substance in a perfect state, and erect immense reefs of it; the molluscous animals cover themselves with it; and lastly, all the classes of inferior beings produce and apply it to use daily.

The oldest lime-stone is found in the primitive formations: it is usually very crystalline and large grained, mostly of a white or grey colour; when in large masses, or in beds alternating with gneiss[a] and the schists, its crystalline appearance seems to diminish, in proportion as it rises to the highest limit of these formations.

The primitive stratified lime-stone frequently contains foreign minerals included in it, such as quartz and mica. This latter substance gives it a schistose aspect, and it then makes it a real calcareo-micaceous schist.

[a] "A compound rock, consisting of felspar, quartz, and mica, disposed in slates, from the predominance of the mica scales. Its structure is called by Werner, granular-slaty. This geognostic formation is always stratified:" &c.—*Ure's Chemical Dictionary.*

The limestone of the intermediate or transition formations, is essentially distinguished from the preceding, by its position. It may otherwise exhibit the same mineralogical characters. We should remark, however, that although crystalline, it is very small grained, and that we seldom meet with any great extent of it that is decidedly of a granular texture; that its fracture is most frequently tabular, and tends to approach the compact; that it generally belongs only to the secondary formations; and lastly, that its colours are in general lively, intense, and agreeably mixed.

In the secondary formations, of which lime-stone constitutes the greater bulk, geologists in France remark two stages or formations: 1st, the Alpine, 2d, that of the Jura. These formations are sometimes separated by beds of sand-stone, gypsum, marl, clay, and rock salt, in conformable beds.

The deep grey Alpine lime-stone, after passing to a black or blueish grey, exhibits a compact tabular fracture, sometimes granular, which admits of its being made use of as marble. The Jura lime-stone which is of a clear grey or yellowish white colour, with a compact and smooth conchoidal fracture, very often takes an oolitic structure.

It is principally in the secondary formations that we meet the argillaceous lime-stone calculated to furnish hydraulic limes. We often find it in beds underlying the Alpine lime-stone, and between this ast and the Jura lime-stone, which is itself frequently alloyed with clay[b].

[b] An abundant supply of hydraulic lime-stone is distributed through England by the lias-beds, which the geological maps represent traversing the island from north to south, commencing at Whitby, on the Yorkshire coast, passing through that county, and the west of Lincolnshire, bordering Nottinghamshire, thence following for some distance the east bank of the Trent, afterwards crossing Warwickshire and Gloucestershire, and making its appearance in various parts of Somersetshire, finally terminating near the junction of the counties of Devon and Dorset, in the neighbourhood of Lyme. " This formation consists of thick argillaceous deposits, constituting the base on which the whole oolitic series reposes. The upper portion of these deposits, including about two thirds of their total depth, consists of beds of a deep blue marl, containing only a few irregular and rubbly lime-stone beds. In the lower portion, the lime-stone

The shelly formations and the marls of the tertiary strata are also very often argillaceous, and help to swell the category of hydraulic lime-stones. (Vide the systems of geology.)

(II.) The following is a method of analysis proposed by M. Berthier, for persons who have some acquaintance with chemistry, as well adapted to distinguish hydraulic lime-stone.

" We pound the mineral, and pass the powder through a hair sieve; 10 grammes (154.4 grs.) of this powder are put into a capsule, and we pour upon it, little by little, muriatic acid, (in case we have no muriatic, we may use nitric acid, or vinegar,) diluted with a small portion of water, stirring it continually with a glass rod or strip of wood; we discontinue adding the acid as soon as the effervescence ceases. We then evaporate the solution by a gentle heat, until the whole be brought to a pasty condition; the pasty mass is now

beds increase in frequency, and assume the peculiar aspect which characterises the lias, presenting a series of thin stony beds separated by narrow argillaceous partings; so that quarries of this rock at a distance assume a striped and riband-like appearance; in the lower beds of this lime-stone, the argillaceous partings often become very slight and almost disappear, as may be seen in the lias tract of South Wales: beds of blue marl with irregular calcareous masses, generally separate these strata from the red marl belonging to the subjacent new red sand-stone formation. The lime-stone beds, towards their centre, where most free from external mixture, contain more than 90 per cent of carbonate of lime; the residuum has never been distinctly analysed, but appears to consist of alumine and iron, and in some varieties traces of silex have been found: towards the edges of the beds, however, where they come in contact with the alternating strata of clay, the proportion of alumine is, as might be expected, more considerable. This lime-stone is particularly characterised by its dull earthy aspect, and large conchoidal fracture; in colour it varies in different beds from light slate blue, or smoke grey, to white: the former varieties usually constituting the upper; the latter, the lower portions of the formation. The blue lias, which contains much iron, affords a strong lime, distinguished by its property of setting under water." " The irregular beds consist of fibrous lime-stone and *cement* stones (septaria) so called, because used in making Parker's cement." *Conybeare and Phillips's Geology of England and Wales.*

Towards the southern extremity of the lias formation above referred to, the bed widens in extent, so as to include a small part of Glamorganshire, whence the celebrated Aberthaw lime-stone is procured. But of late years the valuable properties of this lime have acquired it a reputation at other

diffused in about half a litre [c] of water, and we filter it. The clay remains on the filter. This substance is now dried in the sun, or before the fire, and weighed. Or what is still better, it may be calcined to redness in an earthenware or metal crucible, previous to weighing. Very clear lime-water is poured into the solution, as long as it continues to form a precipitate [d]. This precipitate, which is magnesia, is collected as hastily as possible on a filter, it is washed in pure water, dried as perfectly as possible, and lastly weighed."

The weight of the clay, as compared with that of the calcareous mineral dissolved, gives an approximate indication of the place which it ought to occupy in the scale of hydraulic lime-stones. It is important to remark, that instead of clay, there may very probably remain after the first filtration, nothing but a very fine sand, or perhaps a mixture of very fine sand and clay. In the first case, the mineral under examination is incapable of affording any thing but a "poor lime." In the second, we must separate the sand and clay by washing and decantation, and estimate the weight of each independently.

The varieties in the quality of lime did not escape the observation of the ancient architects. Hence they took much

points of the line above traced, as for instance, at Barrow-on-Soar, in Leicestershire, and Watchet, in Somersetshire, both of which will be observed to be within the limit of the same formation. It appears, however, that the quality of the lime is not uniform throughout the whole extent of its range, being in some places slightly less energetic than others, a circumstance no doubt to be attributed to variations in the quality, or proportion, of the alloy contained by it.—Tr.

[c] A litre is equal to about a quart English.—Tr.

[d] It would be better to ascertain by an independent experiment, (viz. by dissolving a separate portion of the mineral in the acid, and adding an excess of lime-water,) whether it contain magnesia or not; because, should there not be any, the lime may be precipitated by carbonate of potash, and its quantity estimated, as a check to the accuracy of the rest of the process. The weight of the precipitate obtained by the carbonate of potash, (which is carbonate of lime, with perhaps a little iron, &c.,) after being washed and dried, ought, with that of the silex and alumina, to make up the whole amount of the mineral dissolved for analysis.—Tr.

trouble to discover relations between these qualities and the external characters of the lime-stones. Their labours were necessarily fruitless, since the chemical composition on which their intimate qualities depend, cannot be exhibited by the weight, hardness, fracture, colour, or, in a word, by what constitutes the physiognomy of a body.

The modern works which have taken up this question, moreover, all repeat, in a pretty uniform manner, the rules which Vitruvius has transmitted to us. The Roman architect had, as he himself confesses, consulted the Greek works which existed in his time; so that all these doctrines seem to refer themselves to a common and very ancient origin; possibly they were perfectly suited to the materials of the countries where the first observations were made.

(III.) The field of our early experiments was singularly enlarged by the researches on hydraulic limes which we undertook in 1824, 1825, and 1826, by order of the Director General of Roads and Bridges, on the lines of navigation from Nantes to St. Malo, from Nantes to Brest, from Briare to Digoin, from Décise to the course of the Yonne near Chatillon, and in the environs of Besançon. We were enabled to examine and to submit to uniform methods of comparison eighty-three varieties of lime, both hydraulic as well as slightly and eminently hydraulic. The details of this examination and of the results which accompanied it, have been confirmed, as regards the canals of Brittany, by M. Bouessel, inspector of division; Piou, Engineer-in-chief; Mequin, Leguay, and Coiquand, Engineers in ordinary, attached to the works of the canal of Isle and Rance; and, for the canals of the Loire and in Nivernais, by Messrs. Vigoreux and Tibord, Engineers in chief, and Barrande, Engineer in ordinary, attached to the service of the canal of the Loire.

(IV.) A very small quantity of iron is sufficient to change the ordinary whiteness of lime, and communicate to it the yellow, red, or greenish-yellow tinge, which we frequently see. It is therefore quite plain, that a lime may be coloured without ceasing to be "rich," while at the same time there is nothing to prevent a very white lime being powerfully hydraulic[e], since it may owe this latter property entirely to the presence of a pure clay, that is to say, one wholly composed of silica and alumina.

(V.)—*The following account of the process alluded to, is extracted from the "Gleanings in Science," vol. i. p. 47.

"The method of analysis was rather different from that usually employed. The lime-stone in powder, was exposed in small covered cupels to the regulated heat of an assay furnace. When the first heating was not sufficient to expel all the carbonic acid, they were again submitted to the fire; and to prove that all the gas had been driven off, a few of the samples were further heated without loss of weight in a forge.

"A very ingenious method was adopted to prove the correctness of this mode of analysis. The lime, rendered caustic by the preceding operation, was converted into a hydrate. The increase of weight was found to correspond with the carbonic acid driven off[f]

"The employment of this method of verification has afforded a very valuable hint towards the solution of the very difficult problem of separating lime from magnesia, or rather

[e] From experiments on various limes, Mr. Smeaton concluded, that their goodness for hydraulic purposes could not be inferred from their colour *previous* to burning, inasmuch as he found blue, whitish, and brown of equally good quality, but they all seemed to agree in falling, after calcination, into a powder of a buff-coloured tinge, and in containing a considerable quantity of clay.—*Construction of the Eddystone Lighthouse.*

[f] Should the mineral consist of carbonate of lime *only*, then the quantity of carbonic acid expelled will be $\frac{22}{50}$ths of its weight, the remaining $\frac{28}{50}$ths being pure lime. This remainder gains an increase of $\frac{9}{28}$ths by slaking, which is the weight of the water which unites with the lime by that process. Hence the following rule for ascertaining the quantity of lime in a compound of lime and magnesia, by the process of hydration. Multiply the increase of weight after slaking by $\frac{28}{9}$ (or by $3\frac{1}{9}$) which will give the amount of *pure* lime, or by $5\frac{5}{9}$ for the quantity of *carbonate of lime* corresponding to it which existed in the mineral previous to calcination. The remainder will be carbonate of magnesia, unless there be other ingredients, which must be ascertained by a separate process. In carbonate of magnesia $\frac{20}{42}$dths or nearly 47 per cent. is magnesia, and the remaining 53 carbonic acid.—Tr.

of estimating their respective quantities. It appears that magnesia does not form a hydrate [g], so that the increase of weight is an index to the quantity of lime in any magnesian lime-stone." In recording some further experiments in prosecution of this subject, the author says, "the results were in every way coincident and satisfactory, and leave no doubt in my mind, that the dry analysis of mixtures of lime and magnesia is capable of much greater precision than the humid analysis, in which according to Daubeny, five, and even ten per cent. of difference will result from the use of different precipitants.

"With care in the process of calcination, and the check operation of slaking, I do not imagine that one per cent. of error should find its way into the result. And the method is applicable even when there is silex or any of the inalterable earths united to the carbonates; provided that the proportion of these be first ascertained by solution in an acid."—TR.

(VI.)—*Extensive beds of the native carbonate of magnesia have been recently discovered in the South of India, near Salem and Trichinopoly, in the Madras Presidency, and the supply is so abundant, that measures have already been taken for turning this valuable material to account as a cement, for which purpose the experiments of Colonel Pasley have shown it to be admirably adapted. Some varieties of the mineral are so

[g] This is not correct, as magnesia does combine with water, and form a hydrate, but not with sufficient rapidity to interfere with the correctness of this method of analysis, if its completion be not unduly retarded. I found by experiment, that a hundred grains of fresh calcined magnesia, powdered while warm, and immersed in water, had in twenty hours gained $18\frac{1}{2}$ grains in weight, and after three days and a half immersion had combined with very nearly the whole of their saturating dose of water, (the weight being then 142.2, and some allowance for loss being necessary, owing to its having been dried and weighed repeatedly) and become converted into a true hydrate of magnesia. It is remarkable also, that as long as this reabsorption of water continues, the mass has a great tendency to *set* and harden, and requires to be continually stirred to prevent its doing so; but as soon as the saturation is complete, the powder becomes quite inert, and incapable of setting under water.—TR.

hard, previous to calcination, as to strike fire with steel, others are of a softer description. The constituents of a specimen analysed by my friend Dr. Malcolmson, of the Madras medical service, to whose kindness I am indebted for the following particulars, are, in a hundred grains, magnesia 47½, carbonic acid 51½, water ½, and silica ½, the proportion of the last ingredient varying in different samples within very narrow limits. After calcination, the magnesia does not slake like lime, but when powdered and made into a paste, a sensible heat is extricated; it is capable of hardening under water, though it is preferable to allow it to dry for twelve hours or more previous to immersion. In time it acquires a firm consistency, and even as a common stucco has been described as of "extreme hardness." An admixture of not more than one and a half times its bulk of clean sand is found to improve its qualities for general purposes, at the same time that it diminishes its cost, but the proportion of such alloy must, of course, be regulated by the use for which the mixture is intended. As a stucco it is considered the most beautiful of all the cements, and that even at Madras, where the chunam, so long celebrated, is made in the greatest perfection. In fact, the only impediments to its exclusive adoption seem to be the cost of transporting it from the situation in which it is found, and the difficulty of preserving its properties, after calcination, unimpaired, it being subject to deterioration by the absorption of moisture from the atmosphere; together with the cost of pulverising it previous to use. With regard to the use of magnesian lime-stone, I shall merely add to what is stated on this subject in the note, that I have met with very excellent common mortar of great hardness which, from the proportions of lime and magnesia which it contained, appeared to have been made from dolomite. And the following quotation from Phillip's Mineralogy, p. 166, where in speaking of magnesian lime-stone, he says, "The lime obtained from it is greatly esteemed for cements, being less subject to decay, owing to its absorbing less carbonic acid from the atmosphere than the lime of common lime-stone," seems to correspond with the opinion I have been inclined to form of it.

I have recently had an opportunity of making some experiments with the calcined mineral referred to in the preceding paragraph. I found it exceedingly hard, and difficult to reduce to a sufficiently minute state of division, but after this was effected, a mass which had been squeezed into the bottom of a jar, and immediately immersed, set perfectly in thirty-eight hours, and gradually hardened; but a ball of the same paste fell to pieces in half an hour. It appeared that in the first case a thin superficial layer of the cement was deteriorated by the action of the water, and became detached from the rest of the mass, but as it could not leave its position, (the surface of the paste being horizontal,) it served as a protection to the parts underneath. In the ball, on the contrary, similar deteriorated parts at the sides and underneath, having no support, fell from it, and exposed fresh surfaces to the action of the water, which were again destroyed by it, till the whole had fallen to pieces. In the mass which set there was a thin layer of this powder covering the surface, which was of a harsh meagre description, altogether devoid of consistency. The cement beneath was in a fortnight already very hard, so that no mark whatever could be made upon it by the nail. It split the vessel (which was a preserve jar of $2\frac{1}{2}$ inches diameter) in which it set, and the crack very slowly, but perceptibly, widened day by day, being at the end of the fortnight about $\frac{1}{40}$th of an inch in width. A quantity of the powder which had been diffused in water and allowed to settle, and had remained undisturbed during fourteen days, was found to have some consistency, and seemed inclined to set. A thick fluid pulp which had been poured into a phial so as to fill it, and tightly corked up, had solidified, and was become nearly dry, but its hardness was not then of the kind that is acquired by immersion, or by drying in the open air, and was more like that of mortar which has dried in the heap. It seemed to want the previous condensation.—Tr.

Since the above was sent to press I have met with the notice of a paper on the sole efficacy of magnesia in rendering certain lime-stones hydraulic, submitted by M. Vicat to the Royal Academy of Sciences at Paris, and which is thus referred to:—"This paper has for its object the correction of an opinion given by M. Berthier in the Journal des Mines of 1832, that magnesia alone has no more efficacy than alumina to render lime hydraulic; from which it would follow that silex was the only essential principle in all cases."

M. Vicat was, for a long time, of the same opinion, which he now declares incorrect; and says that magnesia alone, when in sufficient quantity, will render pure lime hydraulic. He does not explain the degree of energy of these new species of lime, but only affirms that they will solidify from the sixth to the eighth day, and continue to harden in the same manner as ordinary hydraulic lime.

Until his experiments are further advanced, he states that the proportions of magnesia, taken and weighed after calcination, should be from 30 to 40 for every 40 of pure anhydrous lime. The native lime-stones examined and cited by M. Berthier contained only from 20 to 26 of magnesia for every 78 to 60 of lime. It is probable that this want of proper proportions was the cause of his negative results. M. Vicat, in conclusion, points out the importance of these observations; hydraulic lime never having been found in the calcareous formation below the lias, is because the dolomites have never been examined, but it is now probable it may be found in this lower formation."—Vide Journal Frank. Inst. Vol. xviii. No. 4.—Tr.

(VII.) It was for a long time thought, according to Bergmann, that the virtue of hydraulic lime depended upon the presence of a few hundredths of manganese. Guyton was of this opinion. Smeaton, the English Engineer, had remarked, earlier than 1756, that the limes of England fitted for subaqueous constructions, all left an argillaceous sandy residuum after solution in nitric acid. (Construction of the Eddystone

Lighthouse, extracted from the "Bibliothèque Brittanique.") [h] Thirty years after, Saussure announced that the lime of Chamouni, although destitute of manganese, hardened under water, whence he inferred, with reason, that that property depended solely upon the clay. (Tour in the Alps by Saussure.) Lastly, Colets-Descotils, on analysing the compact marl of Senonches, in 1813, found in it very nearly a fourth of silica, a circumstance which led him to conclude that "the cause of the phenomenon was due to the presence of a large quantity of siliceous matter, disseminated in very fine particles throughout the texture of the mineral."

The opinion of Descotils did not invalidate that of Saussure, inasmuch as clay generally contains more silica than alumina; and besides, the two chemists perfectly coincided in the opinion, that the oxide of manganese, if not useless, was at least not an essential element. Such was the state of the question in 1813. To put an end to all these doubts, we determined at this time to proceed synthetically, and to compound hydraulic limes at once, by calcining different mixtures of common lime slaked spontaneously, and clay: the success exceeded our expectations. All rich limes, soft to the touch, afforded the same result. Our experiments, repeated at Paris in 1817, with the lime of Clayes and Champigny, and the clay of Vanvres, soon afterwards in England by M. St. Leger, at Nemours by M. Giraut, and in Russia by Colonel Raucourt de Charleville, coincide with one another without exception.

M. Berthier, Engineer in chief of mines, has since thrown much light on these phenomena, by studying the action of lime, in the dry way, on the oxides, silica, alumina, magnesia, and the peroxides of iron and manganese, taken singly, and afterwards two and two, &c., &c. The results

[h] The Aberthaw lime-stone left a muddy residuum, which being brought into an argillaceous state, was very tough and tenacious while soft; and when sufficiently hardened, being worked into a little ball, and dried in that state, appeared to be a very fine compact blue clay, weighing nearly one-eighth of the original mass.—*Ibid.* p. 107.

at which this able chemist has arrived, are comprehended in the statement preceding the number of this note (in Chapter I.). The reader who is anxious to make himself acquainted with them in all their details, will find them in the 22d volume of the "Annales de Chimie et de Physique," page 62.

NOTES ON CHAPTER II.

(VIII.) The effects of calcination are not confined to driving off the water of crystallization and carbonic acid from compound limestones; it further modifies the constituent oxides, one by the other. In fact, if we treat an argillaceous carbonate of lime by a weak acid, it forms a deposit or insoluble residuum more or less abundant. After calcination, on the contrary, a complete solution is effected; the clay therefore has entered into combination with the lime.

The arenaceous lime-stones still leave an insoluble residue, when we submit them to the action of acids after calcination; and this deposit is more nearly equivalent to what we should get on treating the carbonate, in proportion as the size of the quartzose grains is greater.

It follows from this, that when the silica, in lieu of being minutely divided, as in clay, is on the contrary, dispersed through the calcareous fabric in the form of sand, its cohesion cannot then be overcome, or at least only in an imperfect degree.

Descotils, to whom a part of these observations is due, justly concluded, that silica does not assist to render lime hydraulic, except in so far as its extreme attenuation allows it to be acted upon by the lime, and enter into combination with it in the dry way. From this we may understand, how there may be limes which are not hydraulic, although containing a notable quantity of silica.

(IX.) Dalton has observed that a current of aqueous vapour

accelerates the reduction of lime-stone into lime[i]. This remark ought not to be passed over, in reference to its economy of fuel. We should do well then, if when burning during the dry season, we were to sprinkle the lime-stone with water before charging the kiln with it, and from time to time throw a few buckets of water on the burning faggots near the eye of the kiln. Or, if we are burning coal, to place a large vessel filled with water in the draught of air at the mouth of the vault. We are informed that Lord Stanhope has made a successful application of this principle in England.

(X.) The carbonate of lime, when violently heated in a close vessel, melts, and afterwards crystallises on cooling, arranging itself anew in the state of carbonate; this observation is due to the chemists Hall and Watt; it is more than half a century old.

The carbonate of lime alloyed with clay, when burnt in contact with charcoal, does not become converted into nearly so good an hydraulic lime, as when it is burnt with coal. And the fire from coal, in turn, does not answer so well as the blaze resulting from the combustion of wood or furze. An authentic experiment made at Nevers in 1825, has proved, that charcoal can deprive hydraulic lime of a half of the

[i] M. Gay-Lussac, (Annales de Chimie, No. 63,) has recently proved this by the following interesting experiment. He filled a porcelain tube with fragments of marble, and attached a retort containing water to one end of it. Heat was applied to the tube till the carbonic acid began to be disengaged from the marble, and it was then slackened to a dull red, when the gas ceased to come over. On now heating the water in the retort, and allowing the vapour to pass through the tube containing the ignited fragments of carbonate of lime, a copious extrication of gas immediately ensued, which ceased the moment the supply of aqueous vapour was cut off, and was again disengaged on its renewal. By a similar experiment, M. Gay-Lussac found that the same effect was produced by a current of atmospheric air, whence he concluded, that the effects in each case were to be attributed to mere mechanical agency. But the last experiment would, I have reason to think, have failed of success, had the air been previously freed from moisture; as Dr. Faraday has long since proved, that when the aerial current has been carefully dried by passing it through sulphuric acid, carbonate of lime may be exposed to the heat of the oxyhydrogen blow pipe, without material loss of carbonic acid. As

energy which it would have acquired, had it been burnt with the common flame heat. Colonel Raucourt, the Engineer, has on his part observed, that a mixture of pure lime and clay, burnt at one time on an iron plate heated to redness, and at another in a kiln, constantly gave good hydraulic lime; whilst the same mixture, if burnt with charcoal, merely produced a poor lime. These facts not only establish the utility of the contact of the atmosphere in burning argillaceous lime-stone, but they lead to the belief that an absorption of oxygen may help to exalt those qualities, which hydraulic lime owes essentially to its combination with clay[k]. This opinion however is not, after all, supported by any fact.

(XI.) Different degrees of calcination may cause lime-stones to lose from 10, 15, 20, to 40 per cent. of their weight. Thus we have various subcarbonates with excess of base, some of which slake in water, and some not[l]: these last possess some remarkable properties which we shall explain. Having divided a certain quantity of pulverized chalk into ten equal parts, we spread the first portion upon an iron plate heated to a cherry red, where it underwent a calcination of three minutes: the second was left six minutes, the third nine, and

no mention is made in the account of M. Gay-Lussac's experiment, of this precaution having been attended to, it is very probable that it may have been omitted, and that this circumstance has led the distinguished French chemist to adopt a theory of the action of the vapour, which seems to be quite inconsistent with the fact above stated.—Tr.

[k] On the other hand, Colonel Pasley remarks, that coal dust, (which is a *de*-oxidating substance,) though not absolutely necessary to all artificial cements, does no harm to any; and his experiments proved that it often did good. A proportion of about one-twentieth of the compound was mixed previous to burning, the object being to restore the iron of the clay as much as possible to the state of protoxide, during the calcination. (Observations on water cements, p. 6.) The clay used in his experiments was of a blue colour, very soft and fine, and was obtained from the Medway at low water, by digging from one to two feet below the surface of the mud.—Tr.

[l] Mr. Higgins found that lime-stone or chalk heated to redness in a covered crucible, or in a perforated crucible through which the air circulates freely, loses only about one-fourth of its weight, however long this heat be continued. (Vide Experiments on calcareous Cements, p. 5.)—Tr.

so on to the tenth, which consequently had to remain for half an hour. During each operation the powder was well stirred in every direction, to assist the equal distribution of the heat. The ten portions so calcined were then kneaded up with a little water, and brought each to a stiff paste of the very same degree of consistency. During this manipulation, not the least sign of effervescence or expansion was exhibited. The first numbers merely exhaled the common odour of moistened chalk; the latter ones exhaled in addition the alkaline smell peculiar to lime, and gave pretty decided indications of causticity.

After two hours of immersion, all the numbers, excepting the first, had set; and after four days, it was altogether impossible to thrust the finger into them. In a word, the first evidence of this solidification, was so similar to that of the hydraulic limes, that it was impossible not to mistake it for it[m]. But the sequel did not justify the hopes which we had a right to form; it is for this reason that we expressly give warning against the trial of lime-stones by a calcination effected on a small scale, that is to say, with small portions of material subjected to a chimney or stove heat.

During the time that we were engaged upon these experiments, M. Minard, Engineer in chief of roads, observed on his part analogous phenomena; but, being without doubt deceived by the rapidity of the first set, this Engineer prematurely announced, that a properly regulated calcination could be made to bestow upon nearly all pure calcareous substances, the property of furnishing at pleasure, either rich lime, or natural cement analogous to Parker's cement. Unfortunately the sequel has failed to confirm these predictions. Our own experiments, as well as those which M. Berthier has since made known, have completely invalidated them.

[m] It is remarkable also, that rich limes when partially regenerated by the carbonic acid of the atmosphere, appear to possess similar properties. Col. Raucourt de Charleville found that common rich lime, after falling spontaneously, by exposure to the air under a shed, and being stirred from time to time to equalize its action, had acquired hydraulic properties in a month's time, so as to be capable of setting under water when used either alone or with sand.—Tr.

M. Berthier is of opinion, that when we throw water upon the sub-carbonates, they are changed into hydro-carbonates, or combinations of the hydrate, and the ordinary carbonate of lime; combinations which are feeble, and of little permanency in the water[n].

The sub-carbonates with excess of base, are much more difficult than the neutral carbonates to reduce completely into lime by a second calcination; a circumstance which accounts for the old opinion common to all the master lime-burners, viz. that it is impossible, let you burn a whole forest, to transform into lime a lump of the stone, which has been chilled before the completion of the calcination.

M. Lacordaire, Engineer of bridges and roads, employed on the works at thejunction of the Burgundy canal at Pouilly, having repeated the experiments made by M. Minard and myself on the pure lime-stones, upon the argillaceous ones of Auxois, has discovered that the latter, when in the state of sub-carbonates, become real natural cements. This observation has led to a number of happy applications to the advantage of the works which he is directing. Thus he has reduced the continuance of the heat, which is usually for six to eight days in ordinary kilns, to three, being confident in his power of turning to account the unburnt lumps which were necessarily formed by this diminished consumption of fuel. To effect this, he slaked the lime by immersion, separated from it the sub-carbonized parts, which he afterwards reduced mechanically to powder, and incorporated with the mortar.

By a singular coincidence of circumstances, the quantities

[n] Some specimens of mortar which I examined in the course of my experiments, and in which the lime was perfectly unalterable by the action of tests, (being completely neutralized,) exhibited a large deficiency of carbonic acid, although they were in a state of such entire disaggregation, as to leave hardly room for the supposition that a union between the silica and the lime, (vide art. 309,) could have taken place. In these there was a considerable quantity of *water* in chemical combination, and it seemed to me probable, that a hydrated sub-carbonate of lime had been formed, together with the true carbonate, with which it was mechanically mixed; and it is remarkable that the mortars whose analyses led me to suspect the existence of the hydro-subcarbonate, were some which had been prematurely ruined by exposure to *damp*.—Tr.

of quick lime and sub-carbonate furnished by each charge, were found to be in the respective proportions (from 12 to 15 cubic metres of the sub-carbonate, to 48 or 45 cubic metres of lime) which experience demands for the production of the best possible hydraulic mortar (of mortars of this kind,) with the addition of a natural calcareous sand found in the country. The economy afforded by the manufacture of this new kind of mortar, cannot be looked upon as a general result. It depends essentially upon the imperfection of the lime-kilns in the neighbourhood of Pouilly, the dearness of fuel, and the scarcity of sand. But let the price of wood or coal be taken at the average rate which it bears every where else, and the form of the kiln be modified in such a way, as to do away with the necessity of expending as much caloric in burning the last fourth or fifth of the charge of stone, as is required for first 3-4ths or 4-5ths, then the method of M. Lacordaire will maintain none of its advantages. The mortar of sub-carbonate of lime does not seem as yet to have been exposed to the action of the weather. It is impossible to form a judgment of its actual qualities, until varied experiments have sanctioned its use on all occasions.

(XII.) At the Monsieur canal (a field kiln is used), seventy-five cubic metres of lime required, on an average, 26 cordes of wood (oak). The corde measures 4.74 steres, which makes 1.64 st. per cubic metre[o].

At the bridge of Souillac, (the kiln is cylindrical, surmounted by a cone,) a hundred metres of lime required, on an average, 107 steres of wood (oak), which is 1.70 st. per metre.

At the Monsieur canal, (the kiln was a field one) forty-seven cubic metres of lime required, on an average, 7000 faggots, measuring together 1050 cubic metres, which comes to 2.234 st. per metre.

[o] The reader will recollect, that the stere (which is a measure only applied to wood, faggots, &c., &c.) is *equal* in capacity to the cubic metre; hence the relative proportions laid down in this and the succeeding para-

At Saint Brieux, (the kiln ovoidal,) fifty cubic metres of lime required 2500 fascines of broom, measuring together 1500 st., which makes 30 steres per metre.

At Cartravers (the kiln is ovoidal), thirty cubic metres of lime take, on an average, 15,000 fascines of broom, measuring together 900 st., which makes 30 steres per metre.

In the coal kilns by slow heat, at Cahors, twelve cubic metres of lime consume, on an average, 4 cubic metres of coal. (The kilns are inverted cones.) This comes to 0.33 m. per metre.

In the coal kilns, by slow heat, at Doué (Maine et Loire), seventy-eight cubic metres of lime consume, on an average, 29 m. 80 c. of coal (the kilns are ovoidal), which comes to 0 m. 36 c. per metre.

The broom and the faggots were measured in large piles, and consequently, with the bulk which the pressure they were then subject to, gave them.

(XIII.) We had occasion, in 1824, to visit the coal kilns established at Paris, near the bridge of Jena, for burning the artificial hydraulic lime. These kilns, which were built of the shape of inverted cones, after the old method, worked badly. The quantity of "core," or unburnt lime was constantly so great, that it could not be exposed for sale without previously picking out the worst parts. We again visited these same kilns in 1826, and they worked very well. No change had been made, either in the shape, or capacity of the cones; the improvement was owing—1st, To their not lowering the mass at each drawing of the charge more than from 30 to 35 centimetres (11.83 to 13.78 inches, Tr.), instead of 0.64 m. (25.2 in., Tr.), the limit used before. 2dly, To their finishing the charge at the level of the top of the kiln, in lieu of heaping it up in a pyramid as was done before. 3d, and lastly, To the use of a peculiar mixture of coal and coke, and to a well regulated and uniform distribution of this mixed combustible.

graphs, may be applied with equal correctness to any other kind of measure: for instance, in the present case, a cubic *yard* of lime would require 1.64 yards of wood.—Tr.

(XIV.) The ovoidal kilns are subject to the same caprices as the conical ones. They have been seen to work perfectly well for a long period, and then become suddenly out of order, without any apparent cause.

The lime furnished by the kiln whose section is given in fig. 9 (plate 1), is always either too much or too little burnt. A current of flame, probably owing to the sudden contraction of the upper opening, rises with force by the circumference, while the middle burns badly.

The kiln, fig. 10, has been a long time in use, it works tolerably; but Nos. 11 and 12, in the last section, answer the best. Number 11 produces 6.84 cubic metres (241½ cubic feet, Tr.), and No. 12—4.50 cubic metres (158.8 cubic feet, Tr.) of lime per day.

It is in the ovoidal kiln, that Messrs. Debline and Donop propose to burn lime with a peat fire. They affirm that two steres of that material, which do not cost more than half the price of wood, are sufficient to burn a cubic metre of lime; there is therefore much benefit to be derived from using it, wherever it is provided by nature. (The Society of Encouragement have decreed to Messrs. Debline and Donop the prize which it had proposed for the invention of an economical mode of burning, applicable to the manufacture of lime.) The turf is placed upon a grate fixed beneath the vault of the charge, and burns with a flame, in the manner of wood.

(XV.) In the flame kilns from five to seven metres (16.4 to 23 feet, Tr.) high, it is extremely difficult, or rather impossible, to burn the upper layers properly, without exceeding the measure of right calcination for the lower strata; this inconvenience is of little moment with respect to rich lime; but it is accompanied by serious consequences with the argillaceous lime-stone, which in such case vitrifies, and is no longer good for any thing.

We have proposed to the Administration of Roads and Bridges, a kiln, and a mode of calcination, which would appear to obviate the inconveniences above mentioned, at least in a great measure. The following is a description of it.

The form of the lower part of its interior, to the height of about two metres ($6\frac{1}{2}$ feet, Tr.), is cylindrical, on a circular, or elliptic base (figs. 5 and 6, plate 1). The remainder, to the height of 5 metres (16.4 feet, Tr.), is finished off by a conical hood truncated at its vertex, so as to leave an opening of 0 m. 60 (23.6 in., Tr.) for the smoke and heated air. The inferior capacity is divided by partitions into two or three rounded chambers. In the first case, the partitions are raised 3 metres (9 ft. 10 in., Tr.), in the second 2.60 metres ($8\frac{1}{2}$ feet, Tr.) above the floor. The object of these various arrangements is, 1st, to avoid angular parts, in which the calcination always proceeds badly; 2d, to keep up the intensity of the heat in the upper parts, by a more powerful concentration than takes place in prismatic or cylindric spaces whose sides are perpendicular; 3d and lastly, the partitions are intended to be adapted for the alternate calcination of the lower strata, without the discontinuance of it to the upper layers. For instance, supposing the kiln with three chambers to be loaded on vaults, as is usual, with an easily vitrifiable argillaceous limestone. We shut very exactly the openings of the two fires, *b* and *c*. (fig. 6, pl. 1.) We light that in *d*, and keep it up for two days. Before the close of the second day, we unclose the opening of the kiln *b*, and ignite it without delay; as soon as it begins to draw, we imperceptibly slacken that of *d*, which we allow to go out altogether, when *b* is in full activity, by quickly closing the aperture. We change the fire of *b* for that of *c* with the same precautions. It will then by this means ensue, that the whole of the upper, and one portion of the middle region, will have undergone 144 hours of the direct heat, while the lower parts will only have received 48 hours of the same heat.

The kiln with two chambers, is intended for the argillaceous lime-stone of medium hardness; we proceed in the manner above described, and each chamber will receive 72 hours of the direct heat.

The height of the partitions, and the continuance of the heat in each chamber, are evidently two circumstances which depend upon the hardness of the stone, the quantity of clay it contains, and the heat given out by the wood within a given time, according as it may be green or dry. One or two experiments made by an intelligent foreman, ought to be sufficient to decide all these questions. Besides, nothing is easier than to raise or depress the partitions, without in the least injuring the body of the kiln.

Much stress ought to be laid upon the necessity of changing the heat from one fire to the other, without in the slightest degree slackening its continued and general action in the regions situated above the partitions.

When, by reason of its peculiar nature, or its friability, the lime-stone is incapable of bearing the pressure to which the vault and piers of the charge are subjected, that vault is divided into two others, by placing a pier in the middle. Should this precaution appear insufficient, we build the facing of the piers, and the first course of the intrados of the vault, with incombustible stone, or common lime-stone. The latter expedient, in a word, consists in building a skeleton vault of refractory bricks, as is practised in the pottery furnaces.

NOTES ON CHAPTER III.

(XVI.) At the canals of Saint Martin and Saint Maur, the artificial hydraulic limes have been put to every possible proof; they have been applied to the masonry of locks, to the coating of the basins, to subterranean vaults, &c., &c. The following are other instances, in which mortar manufactured with the same ingredients, has fulfilled different functions.

1. At the bridge Duke D'Angoulême, over the Dordogne at Souillac, the foundation of one of the piers is entirely laid on a beton composed of sand, flints, and artificial hydraulic lime. This beton was immersed in five metres (16.4 feet, Tr.) reduced depth, across a current, in a caisson without a bottom. After eight months it bore the weight of 2,500,000

kil. (2462 tons, 7 cwts. 2 qrs. 0½ lb., Tr.) distributed over a surface of eighty metres. (95 sq. yds., 6 ft. 76 in., Tr.)[p].

2. The bridge of Melisey is in the same way founded upon a bed of beton of artificial hydraulic lime.

3. In the repairs of dilapidations, made at the bridges of Tavernay and Baudoncourt, with the same lime made use of at the bridge of Melisey; the beton immersed did not after fifteen days sink more than two millemetres (.08 inch, Tr.) under the pressure of an iron needle a centimetre (.39 inch, Tr.) square, loaded with a weight of 300 kilogrammes (5 cwt. 3 qrs. 17 lbs. 14 oz., Tr.). The experiment was made by M. Lacordaire, Engineer.

4. Experimental pointing with mortar, composed of sand and hydraulic lime, made use of at the base of the Triche mole at Saint Malo, on the 12th July, 1821, by M. Robinot, the engineer. This mortar, which was introduced into joints from which the water flowed as from a spring, had, when we visited the work, resisted the action of the sea for the years 1821, 1822, 1823, 1824. It was in very good condition, although over dosed with lime. Now, there is no Engineer, who is not aware of what the effects of the ocean upon the mortars of revetments are at Saint Malo. M. Bernard, Engineer of bridges and roads, attached to the works at the harbour of Toulon, wrote me word on the 26th October, 1826, that having had occasion to demolish certain portions of masonry, *five or six months' old*, he found that the mortar of artificial lime which had been made use of, had already bound the stones together with such force, that it was often more easy to break them, than to tear them asunder. All Paris could in 1826 have witnessed the hardness which the beton of the floor of the basin at the canal of Saint Martin had acquired. The fragments which were taken out where the floor had sunk, were like mill-stone; it was therefore determined to turn them to use. They were broken by heavy

[p] This is at the rate of about 6402 lbs. per English square foot; or very nearly 45 lbs. per square inch.—Tr.

blows of a mallet, in order to employ them in lumps for the fabrication of a new beton destined to repair the breach. These examples will suffice, we trust, to show the value of those attempts which some persons have made, to cast a slur upon the efficacy of artificial mortars. We have been anxious, in the preceding pages, to rest solely upon examples in the large way. The experiments on a small scale, would not have been sufficiently conclusive; on this account we have passed over in silence, the examination instituted by the Council of Civil Buildings, on the 24th December, 1821, and the detailed report of the Commission nominated for that purpose. From which report it results, that the artificial lime at Paris, is superior to the natural hydraulic lime of Senonches for the first few months of immersion. The experiments of the Commission being incompetent to establish the progressive law of solidification, we ought, in fact, to refrain from coming to any conclusion, as to the future state of the limes compared together.

(XVII.) The efficacy of the oxide of manganese, first supposed by Bergmann, and afterwards assented to by Guyton, suggested to the French chemists, two methods of procuring artificial hydraulic limes; the first, consisted in mixing four parts of grey clay, and six parts of the black oxide of manganese, with ninety parts of lime-stone reduced to powder, and calcining the mixture. The second, was to add to common quick lime, a certain quantity of white iron-stone, which is in great part composed of carbonate of lime and manganese.

These two methods, supposing them to be good, must have been of very limited use; for the peroxide of manganese, and the white iron stone, although abundant natural products, are nevertheless not so much so, as to be used as materials for large buildings.

M. Pach, Professor of the Arts and Sciences at Stockholm, informed me that, in reliance upon M. Bergmann, they had in Sweden constructed a whole lock with mortar of rich lime and the peroxide of manganese, but that the wretched condition of the masonry had necessitated its demolition.

(XVIII.) We shall here relate the particulars of the first attempt at the manufacture which has been made on a large scale. It took place at the works of the bridge of the Duke d'Angoulême, at Souillac.

They had a large stock of rich lime slaked to powder by immersion, and kept in that state under a shed. A grey, slightly effervescent clay of the neighbourhood was procured. This, when dried in the sun, and reduced to fragments of the size of a nut, acquired the property of dissolving and diffusing itself spontaneously in water, forming a pulp, which it was sufficient to stir about a few seconds, to render very fine. To effect the mixture of the materials, a score of little heaps of lime in powder, about one-fourth of a cubic metre (8.8 cubic feet, Tr.), were made on an area of a certain size. Each of these being hollowed in the middle like a funnel, received a fluid pulp, composed of 0 m. 03 c. (1.059 cubic feet, Tr.) of clay and water, in such proportions, that by beating and amalgamating them with the aid of pestles, there resulted a tolerably stiff paste. The partial products of each division were accumulated together into two or three large heaps, which were allowed to acquire such a consistency, as to admit of their being afterwards divided into irregular fragments, by means of shovels, or an adze. These fragments, which were of the size of small blocks (" moellon "), were removed and spread out on the ground, to acquire a proper degree of hardness by drying. They were afterwards taken up, to be housed under a shed appertaining to a lime-kiln, like No. 5, pl. 1 (except the partitions), and of the capacity of 100 cubic metres. (130 cubic yards, 20 feet, Tr.)

The base of the charge of the kiln was composed of common lime-stone, and the upper part, of the artificial lime.

An exact extract from the list of the days' labour, and other items of expense, gives the following detail :—

MATERIALS.

	Francs.
34 m. 55 c. (45 cub. yds., Tr.) rich lime in the caustic state, which will be made by the burthen of the kiln (*v. statement*)	00.00
5 m. 76 c. (7½ cub. yds., Tr.) of clay, measured in powder, at 6 francs[q] per metre	34.56
43 m. 18 c. (56·4 cub. yds., Tr.) of lime-stone of rich lime, to furnish the lime credited; and to form the base of the charge, at 3 f.	129.54
150 m. (196 cub. yds., Tr.) steres of fire-wood, at 4 f. 20 c.	630.00
Carried on	F. 794.10

WORKMANSHIP.

	Francs.
Brought forward	794.10
Slaking 34 m. 55 c. of rich lime by immersion; 56 days' work, at 1 f. 50 c. per day	84.00
Mixing the same lime with the clay; 140 days, at 1 f. 50 c.	210.00
Dividing the paste and putting out to dry in the sun; 65 days, at 1 f. 50 c.	97.50
Lifting the fragments, about 50 cubic metres (65 yds. 9 ft., Tr.), carrying them, and measuring them under the shed; 16 days, at 1 f. 50 c.	24.00
Loading the kiln with 43 m. 18 c. of lime-stone, including building the vaults; and also with 50 cubic metres of artificial lime:—	
7 days, a master burner at 3 f.	21.00
36.40 assistants at 2 f.	72.80
Carried over,	1303.40

[q] As the cost of materials necessarily varies with every locality, I have not thought it necessary to reduce these items to their corresponding values in English money.—TR.

	Francs.
Brought over,	1303.40
Burning and keeping up the fire for six days and six nights, as follows:—	
12 days a master burner, at 3 f. . . .	36.00
24 do. of his assistants, at 2 f.	48.00
	1387.40
Sundries at 20 per cent.	277.50
Cost of 50 cubic metres (65 cubic yds., Tr.) of artificial hydraulic lime, which will be diminished to 40 (52 cubic yds., Tr.) by the shrinkage of the material	1664.90

At this price the metre costs 41 francs 62 centimes. Such is the result of an experiment made under the most unfavourable circumstances, that is to say, without the aid of any previous experience, and without any other power for the manipulation, than manual labour.

At the bridge of Melisey, the cubic metre of twice-burnt artificial hydraulic lime like the above, amounted, inclusive of 20 per cent. for sundries, &c., to 38 francs, taken at the kiln; which is not far from the price which it came to at the bridge of the Duke d'Angoulême[r].

NOTES ON CHAPTER IV.

(XIX.) The vapour which rises during extinction, turns paper tinged with mallow, green, which is occasioned by a portion of the lime in a minute state of division, which the vapour carries along with it.

We introduced successively several kinds of quick lime, still retaining some carbonic acid, into a large earthenware retort, with distilled water; and then by means of a bent

Col. Raucourt de Charleville estimates the cost of the factitious hydraulic lime, twice kilned, at one-third more than the ordinary quicklime of which it is composed.—*Essai*, p. 215.

tube, collected the vapours and gas disengaged during the extinction, in a large bottle full of filtered lime water, inverted over a pneumatic trough filled with the same water. When the operation was completed, we corked and withdrew the bottle, which contained about two-thirds of gas, and one-third of lime water, and then shook it for a long time, without perceiving any precipitate in the liquid; having returned and uncorked the bottle again in the trough, not the least absorption was manifested. Lastly, a lighted match burnt in the gas which it contained, without its light changing in colour or intensity; therefore, there was nothing which escaped from the retort, but common air and water.

(XX.) The masons say of rich lime which slakes to dryness, that it *is scorched*, and has lost its properties; it in fact loses the faculty of producing so fine and so perfect a pulp, as when it passes immediately from the state of quick lime to the state of paste; but it gains in respect to its binding qualities.

(XXI.) This method of extinction is due to M. de Lafaye, who published it in 1777, as a secret restored from the Romans. He founded his opinion upon a forced interpretation of the expressions "lapis calcis intinctus in aquâ," which we find in Vitruvius, Book II. Chap. 5, and "perfundere calcem, perfusio calcis" in Saint Augustin, Book XXI. Chap. 4. of the "City of God." It is sufficient to read attentively the passages from which these words are extracted, to be convinced of the very small probability of the interpretation of M. de Lafaye. His process was very much in vogue at the time. Fleuret, professor of architecture in the old military school, published in 1807, under the title of "L'Art de Composer des Pierres aussi dures que le Caillou," a treatise on the fabrication of mortar by Lafaye's process; which process, however, he modifies by endeavouring to retain and turn to use, the purely aqueous vapour which is disengaged at the moment of effervescence. "This vapour," says Fleuret, (pages 40 and 41,) "arouses and stimulates the appetite of the workmen, whence I conclude, that it contains principles suited to the regeneration of lime, and consequently, to the induration of the mortar." The other reasonings by which the author endeavours to demonstrate the excellence of his method, are pretty much of the same force. Experience has done justice to the exclusive pretensions of M. de Lafaye, by pointing out in what cases, and with what kinds of lime, it is proper to apply the extinction by immersion. It has also afforded an explanation, in the difference of the lime made use of, of the success which M. Fleuret met with at Metz, and the disastrous result of his attempts at the basin of Villette.

	Water absorbed.	Bulk of the paste.
	Kil.	Vol.
(XXII.) 100 kilogrammes of rich lime, converted into a thin paste by the first process, give	291	350 [s]
The same, previously slaked by immersion	172	234
The same, first slaked spontaneously . .	188	258
100 kilogrammes of hydraulic lime, reduced to a thin paste by the first process, give	105	137
The same, previously slaked by immersion	71	127
The same, previously slaked spontaneously	68	100 [t]

On examining these results, we see at once the enormous

[s] By a mean of three experiments, I found that a cubic foot of dry slaked *rich* lime, (prepared from sea shells and slaked by immersion,) when thrown loosely into the measure, and shaken down, but not compressed, and then striked, weighed very nearly thirty pounds. And the experiments of General Treussart show, that rich slaked limes, when converted from the dry powder to the state of *paste*, are reduced to about one half their bulk. These facts will be of use in reference to some of the directions contained in this work, which generally apply to mixtures of sand with lime in the latter condition.—Tr.

[t] As the numbers in this table are merely comparative, I have thought it better not to encumber it, by reducing them to their equivalents in English weight.—Tr.

difference in expansion, which results with the rich limes, from the adoption of the two last methods of extinction. We notice besides, that with the exception of the very white limes, all the others acquire a clearer colour by the ordinary mode of extinction, than when they are subjected to spontaneous extinction, or extinction by immersion. Now we know, that many substances may be blanched, and acquire clearness, by a minute subdivision of their particles, which become so many more faces adapted to reflect white light. Such are the black marbles, green bottle-glass, &c., and which give white powders by pulverisation.

(XXIII.) Let us suppose, that making use of the three common processes of extinction, we have prepared three equal volumes, V′, V″, V‴, of the same degree of consistency, (in thin paste,) with the same rich lime; the elements of V′ being, for instance,

		kil.		kil.	
	Quick lime	100	water	290	
Those of V″	will be	150	do.	253	including some
Ditto V‴	will be	135	do.	254	carbonic acid.

If then to each of these equal volumes V′, V″, V‴, we add for example, the same quantity of sand, we shall obtain three mortars equally rich in appearance, yet, nevertheless, very different in the proportions of lime and water. The consequences of this difference are important. In general, in practice, when we make use of lime slaked by the ordinary process, we never know how much we put into the mortar. In fact, between the limits of the consistency of a good paste, to that of a milky one, the lime may receive from 130 to 400 kil. of water for every 100. There is no point established; all depends upon the pleasure, or rather upon habit, which guides the hand of the workman entrusted with the slaking.

(XXIV.) Rich lime, when slaked in paste, and covered with fresh mould in a trench secure from absorption, may be kept many ages. Leoni Baptiste Alberti speaks, (Book II. Chap. XI.) of having seen lime in an old trench, which had been left for about five hundred years, as many evident indications led them to conjecture; that this lime was still so moist, so well tempered, and so rich, that neither honey, nor beasts' marrow were more so.

(XXV.) Lime, housed in this way, becomes carbonised superficially, it forms itself a crust, not very hard to be sure, but sufficiently so to protect the internal particles. Were the powder to rest upon a damp floor, the lime would suck up the water with force, and would pass to the solid or pasty state, according to its being hydraulic or rich.

(XXVI.) It is this faculty of hardening quickly in a trench, which gives the hydraulic lime so bad a reputation amongst the workmen: so that it is only in the last extremity, and for lack of any other, that they will consent to make use of it. They then endeavour to retard its *set*, by drowning it as much as possible; this is what they call *deadening* it; an expression which is by no means synonymous with *slaking*, but which properly signifies depriving the lime of the power and energy which cause it to harden after slaking. But spite of their efforts, the hydraulic lime always ends by hardening in a certain time. They then determine upon breaking it up with hammers, and working it up to the condition of paste, with the addition of water; with such a lime as this they make their mortar; it is needless to enquire if it be of good quality.

(XXVII.) Extinction by immersion is not, as one might suppose, a difficult operation, more particularly when it is done on the large scale. We shall proceed to describe the method employed at Doué, by Messrs. Ollivier, brothers, contractors for the supply of hydraulic lime for the Nantes and Brest canal, for the passage to Nantes.

On leaving the furnace, the lime is put into a bucket, the bottom of which opens and lets down at pleasure. This bucket is suspended at the end of a rope hooked to the jib of a crane. It is lowered into a basin full of water, and then raised after a few seconds, and a quarter of a turn of the crane brings it above an opening left in the roof of a small vaulted building. The bucket opens by the trigger of a spring, and the lime falls into this building, which they call the magazine of immersion. In it, they thus collect all the lime which leaves the kilns during the day's operations, and next day it is found reduced to powder. The magazine empties itself by an opening into a building below; the lime being pushed through this opening, first falls into an iron meshed cylinder, which by a rotatory movement separates the unburnt lumps, and discharges them outside the building. The powder which escapes through the cylinder, falls into a hopper, by which it is conveyed into a second cylinder of metallic web, which performs the office of a finishing bolter. From this it passes into two hoppers fixed to the summits of two vaults, through which they descend. These vaults support the floor of the sifting house, and form the magazine for delivery, on the ground floor. The shoot of these hoppers descends vertically from the summit of the vaults, and opens or shuts at pleasure, over sacks placed at the end of the beam of a balance.

By these means, more easy to understand than to describe, the lime is slaked, sifted, put into sacks, and weighed, in the most simple and expeditious manner; it is embarked upon the Loire. In this way it may be carried many months, even a year, without experiencing any deterioration.

(XXVIII.) A trial on the large scale, with 60 cubic metres, (78 yds. 12 ft., Tr.) of quick lime, at the works of the Duke d'Angouleme bridge at Souillac, proved the excellence of this process. The lime taken from the heap heated and slaked again very well, after five months of a continually rainy winter.

NOTES ON CHAPTER V.

(XXIX.) Pure lime and water combined, constitute the hydrate of lime. It is obtained, 1st. In the solid state, white, and more or less pulverulent, by exposing lime, in a pulpy state, to the heat of a spirit lamp. 2dly. In a crystallized state, by placing lime-water under a glass receiver with concentrated sulphuric acid beside it [u]; the lime then crystal-

[u] Concentrated sulphuric acid being very liable to deterioration by exposure to the air, in consequence of its powerful avidity for moisture, it would be as well, previous to employing it for the purpose for which it is

lises in regular hexahedrons, composed of 0.70 of lime, and 0.30 water. 3dly. When a cold saturated solution of lime is boiled, the hydrate of lime is precipitated in the form of small crystals.

Such are not the hydrates referred to in Chapter V.; under that denomination we comprehend all the solid combinations or pasty mixtures of water and every kind of lime.

(XXX.)—*By the method described in the note to Article 82, and availing myself of the advantage of being in superintendence of the erection and repairs of numerous public buildings, I have been enabled to undertake and carry through a connected series of observations, as to the rapidity of the absorption referred to in the text, in the case of stuccoes freely exposed; and in order to exhibit this action more clearly, I have thrown together in the form of a table, the mean results of many hundred measurements, showing the average progress of induration of mortars of the common sort. But the individual observations themselves are not in such close accordance as might be expected; the depth of penetration of the carbonic acid of the air, being influenced by smoothness of surface, and compactness of structure, to an extent, which renders it difficult to secure close uniformity in the results.

In referring to this table, however, I ought not to omit to mention, that the whole of the stuccoes from which the measurements were taken, were strictly similar in composition, and were prepared by mixing one measure of a very pure slaked lime, with one measure and a quarter of clear sharp sand, and incorporating them well together by beating. They were also laid on and finished in an exactly similar manner, and the different series were all fully exposed in the same situations. The measurements here given, may therefore be considered to approach the true average, under the conditions above stated, as nearly as possible.

here proposed, to make sure of its energy, by boiling it for a short time. When this cannot be done, and in those cases in which its use may be attended with inconvenience, the dry chloride of calcium in powder may be used as a substitute.—Tr.

Table of the Depth of Penetration of Carbonic Acid into different Mortars, after exposure to the influence of the air for various periods.

TIME OF EXPOSURE.	Mean depth of penetration of the carbonic acid.
	Lines.
1 Week	0.25
2 do.	0.50
3 do.	1.0
1 Month	1.5
2 do.	1.875
3 do.	2.00
4 do.	2.37
5 do.	2.9
6 do.	3.3
7 do.	3.85
8 do.	4.07
9 do.	4.35
10 do.	4.6
11 do.	4.8
12 do.	4.9

Being very nearly half an inch at the end of a year's exposure. A similar set of experiments was commenced with mortars of eminently hydraulic lime and sand, and in every other respect similar to the above, which as far as they went gave very nearly the *same* results. I was, however, compelled to abandon them, before they were sufficiently numerous to exhibit the mean depths for the respective periods with sufficient exactness, (by the mutual compensation of errors) to admit of being associated together in a tabular form.—Tr.

(XXXI.) In plate second we have given representations of the carbonated bands of various sections, cut from prisms of lime and mortar a year old. It would be difficult to credit, did we not see it, how great an obstacle a smoothness of surface presents to the penetration of the carbonic acid.

It is not certain, besides, that the reunion of this acid can ever be complete, even in the most accessible parts. The following are the results of the analyses of some very small pieces of hydrate of lime, which had been exposed to the air for eight years.

		Lime.	Carb. Acid.	Water.
The hydrate of very rich lime gave, for every 1000 parts	When slaked by the ordinary process .	564	424	12
	Ditto ditto by immersion .	561	416	23
	Ditto ditto spontaneously .	567	422	11
				Clay and Water.
The hydrate of slightly hydraulic lime gave, for every 1000 parts. . .	When slaked by the ordinary process . . .	510	311	179
	Ditto ditto by immersion .	570	278	252
	Ditto ditto spontaneously .	517	355	128

Turmeric paper, when applied in a moistened state to the centre of the broken pieces, did not change colour; they had therefore no causticity left, although the proportions of carbonic acid were incomplete[u]. We shall observe, moreover, that the hydrates obtained by the second process of extinc-

[u] This fact, which has been exhibited by the researches of several eminent chemists, is so difficult to account for in a satisfactory manner, that I have taken pains to repeat the analyses of cements exhibiting it, with a view to detect, if possible, any inaccuracy in the process, to which to attribute the apparent deficiency in the quantity of carbonic acid separated. The results, however, of many experiments made with this object, have all tended to confirm the statement contained in the text, however anomalous it may appear to be, in reference to the received opinions regarding atomic combinations. No case has occurred to myself, among analyses of cements of various ages, in which I have found the lime fully saturated, (although entirely deprived of its alkaline qualities,) and some cases have presented themselves, in which the deficiency has been decidedly marked, even after an exposure of considerable duration, as for instance, one-fourth the equivalent dose of acid being deficient after an exposure of a couple of centuries.

Thus we see by referring to the table of analyses of old mortars, by Mr. John (App. LX.), that No. 1, though no less than 600 years old, contained only 5 of carbonic acid to 8.7 of lime, being 73 per cent. of the saturating dose; No. 2, of the same age, contained only 24.3 per cent.; No. 3, 1800 years old, 75 per cent.; No. 4, the same age, 63½ ditto; and No. 5, whose age is not stated, not quite 42 per cent.

As no reference is, however, made in these analyses, to the use of any tests to ascertain the condition in which the lime existed in the cement, rendering it *possible* that part might still have remained unaltered by the action of the air, I subjoin the following extracts from analyses by myself, in which that point was particularly attended to; a part of the specimen operated on being in each case reduced to powder, and minutely examined, to detect the existence of lime in the caustic state; but invariably without success, although I availed myself, in addition to the usual method, by

tion, are those which took the least carbonic acid, and the greatest quantity of water; a result in conformity with what has been stated in Chapter IV., in relation to the unequal capacities of the powders of lime slaked spontaneously, and by immersion, for carbonic acid.

(XXXII.) Vitruvius (Book 2, Chapter II.), and Pliny (Book 36, Chapter XXIII.) speak of a light species of work which the Romans called "Albaria Opera," and into which nothing but lime entered.

Thevenot says (in his collection of reports), that in India

the application of moistened turmeric paper, of the more searching tests described in the Note to Art. 82.

EXTRACTS FROM THE ANALYSES OF VARIOUS OLD CEMENTS, showing the quantities of lime and carbonic acid which they were found to contain.

Number of the analysis in the Table XVII.	Age of the cement analysed.	Weight of lime.	Weight of carbonic acid.	Proportion *per cent.* which the quantity of carbonic acid contained in the cement bears to the full saturating dose.
	Years.			
14	1800	29.1	20.0	87.4
17	45	60.1	39.5	83.6
5	200	99.68	73.0	93.2
3	120	26.	17.0	83.2
6	200	110.9	80.0	91.8
8	200	19.7	12.0	77.5
7	200	110.43	79.0	91.0
10	150	17.0	10.3	77.0
11	400	17.4	11.9	87.0
12	400	13.8	8.0	73.7
13	200	33.9	20.0	75.0
15	Recent.	17.1	10.0	74.4
16	Ditto.	15.6	10.0	81.5

The analyses from which these extracts are made, will be found, at length, in table No. XVII.—TR.

Since writing the above, I have been kindly furnished by my friend Dr. Malcolmson, with the results of some experiments upon the mortar of the pyramid of Cheops, which is described in the Note to Article 268. This cement, though between three and four thousand years old, and exhibiting to the most searching examination not the slightest trace of causticity, yet in one analysis yielded only 5.64 carbonic acid to 11 lime, in 100 grains, and in a second 4.53 carbonic acid to 9.6 lime in the same quantity. In the former case, the carbonic acid amounted to 65.4, and in the latter to only 60 per cent. of the saturating dose. These analyses, which were conducted throughout with great care, and

they plaster the walls with a rough cast of quick lime slaked in milk, and beaten up with sugar, and that they afterwards polish the mortar with an agate. The fact is, that they mix the lime with a little curdled milk, with gingelli oil, and water of jaghery, a coarse very brown sugar, which is derived from the cocoa tree. (Vide the letters of M. de Bruns, inserted at the end of M. de Lafaye's memoirs[x].)

most especially in reference to the separation of the carbonic acid and lime, may be fully depended upon. The particulars of the first of them will be found in App. LXXII.

Some further remarks upon this subject will also be found in the Note to Art. 309.—Tr.

[x] The celebrated Madras chunam is a stucco laid on in three coats, the first a common mixture of shell-lime, and sand, tempered with jaghery water (vide Note to Art. 207.), and about half an inch thick; the second of a finer description, made with sifted shell lime and white fine sand, which is also sifted to free it from pebbles or foreign matter; and this coat, as well as the third, is applied without jaghery, which is omitted on account of its colour, and its frequently containing deliquescent salts. The third and last coat which receives the polish, is prepared with great care; the purest and whitest shells being selected for it, and none but white sand of the finest description, and of that a very small proportion is used, varying from one-fourth to one-sixth. The ingredients of the third coat (as well as the second also, sometimes) are ground with a roller on a granite bed to a perfectly smooth uniform paste, which should have the feel and appearance of white cream. In about every bushel of this paste are mixed the whites of ten or a dozen eggs, half a pound of ghee, (which is butter separated from its caseous parts by melting over a slow fire,) and a quart of tyre, (which is sour curd fresh prepared,) to which some add powdered balapong (or soap-stone) from a quarter to half a pound, which is said to improve the polish. But each master bricklayer has generally a recipe of his own, which he boasts of as superior to all others. The essential ingredients, in addition to the lime and sand, seem to be the albumen (of the eggs), and the oily matter of the clarified butter, for which oil is sometimes substituted. The last coat is laid on exceedingly thin, and before the second is dry; it dries speedily, and is afterwards rubbed with the smooth surface of a piece of the soap-stone (steatite), or agate, to produce the polish, an operation which is sometimes continued for many hours, after which it is necessary to wipe it from time to time with a soft napkin, to remove the water which continues to exude from it for a day or two after completion.—Tr.

NOTES ON CHAPTER VII.

(XXXIII.) The proprietors of the mills on the tributaries of the Loire, between Nevers and Briare, and on the river Isle, in the department of Dordogne, have for time immemorial used, in repairing their works, a mortar composed of the rich lime of the country, and a reddish, very argillaceous pit sand, which they call arène. This mortar hardens more or less in the water, but it is sufficient for the preservation of the hydraulic masonry in which they use it. M. Girard, Engineer of bridges and roads, and in charge of the works connected with the navigation of the river Isle, being naturally struck by so singular a fact, engaged in 1825 in some extremely interesting researches regarding the properties of this kind of sand. The service he was charged with afforded him opportunities of extending his experiments beyond the bounds which confine the mere trials in the laboratory; and the observations which he has made upon mortars of arène employed in large works, such as locks, wears, &c., &c., afford, in consequence, as great a degree of certainty as could be wished for. These observations, which we have eagerly availed ourselves of, have given a place in the domain of science and art henceforth to facts, which, however well they might have been known to the millers of the Isle and others, were not the less hidden and lost to builders.

(XXXIV.) The hydraulic properties of the brown schistose psammites, (grey-wackes of the Germans,) when used in their natural state with rich lime, were discovered in 1824 by M. Avril, the Engineer, when charged with the superintendence of a division of the canal from Nantes to Brest, in the department of Finisterre. "In studying the varieties of schist which compose the soil of our neighbourhood," says this Engineer, (Memoir on the Navigation of the Kergoat, written 1st October, 1824,) "chance threw in my way an arenaceous rock of a yellowish red colour, in a state of decomposition, the action of which, when used as sand in its natural state, and as a cement, after a calcination of ten hours in a lime-kiln, seemed to promise a solution of the problem which I had undertaken. This rock belongs to the species of sandstone called grey-wackes by the Germans, and psammites by M. Brogniard; it is so soft as to fall to powder between the fingers; it hardens in the air and the fire, a circumstance which enables us to build it up into vaults, in the same way as lime, to burn it," &c.

The conversion of the schistose psammites into pouzzolanas by a gentle calcination, is by no means a new phenomenon in the science of mortars, for these substances being essentially composed of silica, alumina, and oxide of iron, form, in some measure, the transition of the schistose rocks into clays, and even seem to differ from these merely in having a lighter contexture, and by the faces of mica which their fracture exhibits. But what truly constitutes a new fact, (and which sooner or later must have led to the discovery of the analogous properties of the arènes,) is the circumstance, that in their natural state they act as pouzzolanas with rich lime; this action it is true is slow, and very variable in its intensity, but it does take place.

(XXXV.) Pouzzolana was in extensive use amongst the Romans[y]. The following is what Vitruvius says of it:—

[y] No particular account being here given of the use of Dutch Tarras, I have extracted and thrown together the following notices of it from Mr. Smeaton's valuable Essay on Water Cements, in his work on the Eddystone Lighthouse.—" I found it in lumps of various sizes, from the bigness of a pea to that of a middle sized turnip: it is of a light greyish, or ash colour, is rather tender than hard, and is very porous, somewhat resembling a pumice stone. There seems to be nothing calcareous in its composition, for aquafortis dropped upon it only wets it like water: to me it much resembles some petrifactions that I have seen; but my more learned friends seem to be of opinion that it is a lava. . . . . . . Although really endowed with those qualities, which have justly obtained it a reputation for water building, yet it is generally admitted to have some properties, that for our use were not quite so eligible. In the first place, though it will cause most kinds of lime to set and become hard under water, as we have seen by several examples, yet if the cement grows dry

"There is found in the neighbourhood of Baye, and the municipal lands lying at the foot of Vesuvius, a kind of powder which produces admirable effects; when mixed with lime and small stones, it has not only the advantage of giving great solidity to common buildings, but possesses the further property of forming masses of masonry, which harden under water. We may account for this property by the great number of burning soils, and hot springs, which give evidence of an extensive subterraneous fire, occasioned by the inflammation of sulphur, alum, or bitumen. Thus the vapour of the fire and the flame passing constantly through these beds of earth renders them light, and converts them into a withered tufa without moisture, so that these three substances (the lime, the stones, and the pouzzolana) modified by the violence of the heat, on being mixed together, form a solid substance as soon as we add water. This mixture quickly acquires so great hardness from the damp which it absorbs, that neither the dashing of waves, nor the action of water, are able to destroy it."

For want of pouzzolana, the Romans made use of pounded brick, which we call cement.

(XXXVI.) Count Chaptal was the first who remarked the unequal manner in which certain of the pouzzolanas of Italy

by a gradual exposure to the air, it never sets into a substance so hard as if the same lime had been mixed with good clean common sand; but is very friable and crumbly; and if, after it has acquired a considerable degree of hardness by immersion in water, it is then exposed to the air, it loses a considerable part of its firmness, and also becomes crumbly; though according to my observation, it never becomes so much so, as if it never had acquired a greater hardness by a submersion in water. . . . In a state between wet and dry, or of being wet and dry by intervals, Tarras is known not to answer well. . . . . Another property of Tarras mortar is, that when kept always wet, and consequently in a state most favourable to its cementing principle, it throws out a substance something like the stony concrescences in caverns of lime-stone strata, called stalactites; which substance from the Tarras comes to a considerable degree of hardness, and in time becomes so extuberant as to deform the face of the walls; and when smoothness and regularity of surface is wanted, as in navigable sluices, mill conduits, &c., it becomes necessary to remove its

and Vivarais behave with sulphuric acid; without deducing therefrom any conclusion with respect to the energy of these substances. The observations which have, in a general manner, established the relations between the qualities of the ingredients of calcareous cements, and their mode of action in respect to acids and lime-water, are entirely our own. They were published by fragments in the different numbers of the "Annales de Chimie et de Physique," from 1819 to 1826. Till that time it was the general opinion of chemists, that clays were more easily acted upon by acids when in their natural state, than after any degree of calcination; a mistake the more fatal, inasmuch as it rendered the supposition of a chemical action taking place, in the phenomenon of the solidification of calcareous cements, altogether improbable.

M. Berthier (in the Journal des Mines, vol. viii. p. 356), has endeavoured to show, in conformity with the opinion of Proust, that the oxide of iron is merely interposed in the ochres, and not in combination with the silica, as Berzelius has stated. This accounts for the facility with which some of the ochreous clays are decolorised by the muriatic and nitric acids.

(XXXVII.) It was Mr. John, of Berlin, who first pointed out the impotency of the most caustic lime on quartz. This able chemist having, in fact, kept a dozen bits of garnet in

roughness by tools; for otherwise the Tarras mortar will grow so much in the joints of these conduits, as to knock off the floats or ladle-boards from the wheels. . . . . A composition of this kind (2 lime, 1 Tarras, 1 sand), is further increased in bulk, by another measure of sand; so that one cube foot or measure of the common Tarras mortar composition will by this means become two, and quite as good in every respect. Finding this idea so far to answer, not only in experiment, but my expectations satisfied in works at large, I was induced to try whether lime would not bear a still greater addition of sand; and I soon found that it would, with good beating, take in, for every two measures of slaked lime, one measure of Tarras, and three of clean sand; which would produce nearly $3\frac{1}{2}$ measures of good water mortar, or full $2\frac{1}{2}$ times the common quantity of mortar from the same quantity of Tarras and lime; this being near upon four times the bulk in solid mortar of the unquenched lime, or the unburnt stone."—Tr.

decoction for eight hours in a fluid pulp made from Carrara marble, these minerals did not lose in it a single atom of their weight; the same was the case with two bits of rock crystal, left for six months in the same fluid.

We have on our part established, as follows, the absence of chemical action by the hydraulic limes upon various kinds of sand.

	Parts.
A certain quantity of white, and entirely quartzose pit sand, was digested in muriatic acid, then washed in much water, and dried at the temperature of boiling water, and weighed	896.000
On the other hand, we took hydraulic lime, in its quick state, fresh from the kiln	300
These substances were made into mortar in the ordinary way in a glass vessel, which weighed 787 parts, and altogether the whole came to 2620, whence it follows that they had taken a quantity of water represented by	647.00
Thus the whole weight of the fresh mortar No. A was	1843.00

A granitic sand mingled with basalt, prepared in the preceding manner, weighed	896.000
Lime as before	300.000
Water absorbed under the same circumstances as above	612.500
Total weight of the mortar No. B	1808.500

These two mortars, when placed in the circumstances most favourable to a chemical action between their component principles, each lost, in fifteen months, 27 per cent. of their weight. Two years after their manufacture, both one and other of them were disaggregated by muriatic acid. The sand of No. A, being separated, washed, and dried, weighed when cool, 892.000 parts. The loss was then only $\frac{1}{223}$ or $\frac{44}{10000}$; it is evidently to be attributed to the second washing. The sand of No. B, treated in the same way, only weighed 883, which gives a loss of $\frac{1}{68}$. We took 500 parts of this same sand, and digested them in muriatic acid; and as there was a disengagement of heat during the disaggregation, we took care to raise the temperature of the acid in the counter-proof in an equal degree, and to continue its action during the same time. The sand, when washed, dried, and weighed as before, gave a deficit of 7 out of the 500, or $\frac{1}{71}$. This result, which differs but slightly from $\frac{1}{68}$, leaves no doubt regarding the cause of the loss exhibited by the sand mixed with basalt; and we may conclude from the above experiments, that the hydraulic lime has no more action on the granitic and basaltic sands, than on those which are purely quartzose.

NOTES ON CHAPTER VIII.

(XXXVIII.) Tile-dust, which has been used in buildings for time immemorial, is evidently, according to our definitions, the most ancient of the *artificial pouzzolanas* known; but the products to which that term was specially applied, previous to the publication of our first researches, do not date farther back than the middle of the 18th century; at which period a Swedish engineer, Baggé of Gothembourg, despairing to obtain the pouzzolanas of Italy for the hydraulic constructions which he was superintending, at a sufficiently moderate price, thought of imitating them by calcining the compact schists, which are found in abundance in Sweden, near Wenesborg.

Count Chaptal, to whom science and the arts are so largely indebted, soon after (1787), made known his experiments upon the calcination of the ochreous clays of Languedoc, and showed, that when properly chosen and prepared, these substances acted precisely in the same manner as the pouzzolanas of Italy. But it was in the number and respective proportions of their principles, that this celebrated chemist endeavoured to find the cause of the phenomena. The consequences of the important observation contained in App. XXXVI. escaped him.

(XXXIX.) The hydraulic virtues of the pouzzolanas, have for a long time been attributed to the presence of iron: our experiments upon the non-ferruginous clays have caused us to abandon that opinion. We should be wrong, however, to conclude, that in the red coloured pouzzolanas, the iron is entirely inert; one thing, however, is certain, that is, that its presence is by no means indispensable, since there are very energetic pouzzolanas which do not contain an atom of it [z].

(XL.) Colonel Raucourt of Charleville, the Engineer, to whom, a few days before his departure for Russia, we communicated the singular properties which clays acquired, when calcined in powder on metallic plates heated to redness, was eager to repeat our experiments in Russia; but he has thought, that the slight degree of calcination which the clay undergoes by this means, was not the sole cause of the phenomenon; the contact of the air appeared to him to exert a notable influence; and this suspicion was changed into assurance, by a series of direct experiments, which led him also to examine the effects of the contact of the air on the burning of artificial hydraulic limes. (See pages 130 and 131, of his treatise on mortars.) The conclusions which he has drawn from the whole of his experiments are, (vide p. 136, of the same treatise,) that there is an absorption of oxygen; he again expresses himself very clearly to that purpose, in the note placed at the foot of page 161. But such an opinion not to be mere conjecture, has need of the support of direct experiments; for it is by no means evident, that such is a necessary consequence of the favourable influence of the contact of the air on the conversion of clays into good pouzzolanas. Whilst engaged with the researches of which this

[z] My experiments also tend to establish the author's opinion on this point, both directly and indirectly, as I have met with clay entirely free from iron, which after calcination formed a highly energetic pouzzolana, and on the other hand, a stiff paste prepared for experiment of rich slaked lime and the washed colcothar of vitriol, (peroxide of iron,) was perfectly soft after seven weeks' immersion.—Tr.

chapter contains a summary, we studied the action of lime in the humid way on silica and alumina, when isolated, and taken in various states of cohesion. Mixtures made in the proportions (by bulk) of 200 parts of these oxides, to 100 of rich lime in paste, gave, after three months of immersion, the following relative resistances under the shock of the experimental needle. (Vide the statement of the manner in which the experiments were made, at the end of this volume.)

	Depth of penetration of the needle.	
	Mill.	Inches, Tr.
1st. With the mixtures of the hydrate of rich lime, and gelatinous silica, simply dried in the air	1.79	0.0704723
2d. The same, with the silica calcined to redness	2.50	0.098425
3d. The same, with the silica separated from various clays by boiling sulphuric acid	2.85	0.1122045
4th. The same, with silica separated from these clays, after they had been slightly calcined	4.16	0.1637792
5th. The same, with silica in an altogether impalpable powder, taken from hyalite, by trituration, and successive washings	Indefinite.	
6th. The mixtures of the hydrate of lime, and alumina in a gelatinous state, and slightly dried in the air	18.17	0.7153
7th. The same, with the alumina slightly calcined	12.86	0.504298
8th. The same, with the alumina strongly calcined	Indefinite.	

These experiments prove, that silica may form a good hydraulic cement with rich lime, without being soluble in acids. It suffices for this purpose, that its cohesion be very much less than that it is possessed of in quartz.

We see further, that the hardness of the cement is the greater, the more nearly the silica approaches the gelatinous state, in which condition it is obtained by the aid of chemical agents.

With regard to alumina, we see, that even in the gelatinous state, although it gives rise to an insoluble compound, yet that it produces a substance which, if not soft, is of a very indifferent consistency; we have moreover ascertained, that the peroxide, and carbonate of iron, display no action whatever under the same circumstances.

It is impossible to avoid recognising the effect of a chemical combination in the cases No. 1 and No. 6[a]; and this ought not to surprise us, since it is known, that the solutions of baryta, strontia, and lime, &c., decompose those of alumina and silica in potash, and that the precipitates are binary compounds, of alumina, and of silica, with one of the substances above named.

These curious facts, for which we are indebted to the celebrated Guyton Morveau, have been since confirmed by Professor d'Obereiner and by John of Berlin. It evidently follows from case No. 6, that a firm solidification is not a necessary consequence of the chemical combination of two soft or pasty substances, brought in contact in the humid way.

With regard to Nos. 2, 3, 4, and 5, although analogy

[a] I have found this observation to be fully confirmed, when applied to certain artificial pouzzolanas mixed with slaked rich lime; (vide Note to Art. 314,) but the explanation cannot be extended so as to embrace all cases of the set of hydraulic limes; (vide Note to Art. 307.) I may here, however, notice an experiment made by General Treussart, which appears to bear additional evidence in favour of the conclusions contained in the text: he says, "In mixing common (not hydraulic) lime in paste and clay together, I met with a singular phenomenon, which I am unable to account for. It is, that if you diffuse clay in water, so as to bring it to the consistency of thin pulp, and after bringing the lime to the same condition, mix the two; no sooner is the union effected, than the compound becomes so stiff, that it is difficult to continue the operation without adding a further considerable quantity of water." This seems to indicate a new arrangement of the particles, consequent upon chemical affinity.—Tr.

would lead us to look upon them also as combinations, yet when we take into consideration the state of cohesion of the silica employed, and the known absence of action of the hydrate of rich lime upon quartz, we ought not to affirm such to be the case. We may nevertheless remark, that it is not correct to compare silica, such as it is in quartz, to such as has just been separated from combination; the first in fact is only acted on by soda or potash with heat, whilst the other readily dissolves in them in the cold, a fact which leads us to presume, with some probability, that lime is not devoid of action upon it.

Nos. 1 and 2, which exhibit as much hardness as the best hydraulic cements, lose all their cohesion by an exposure to the air more or less prolonged.

(XLI.) It has been proposed to make use of the reverberatory furnace[b]; but the powder of the clay, owing to the water which it contains latent within it, undergoes an ebullition the moment it touches the incandescent surface, and this raises and dissipates it in clouds; besides, it is no easy

[b] Colonel Raucourt de Charleville describes an air furnace for this purpose, the chimney of which instead of being vertical is curved semicircularly, the lower end of it being horizontal, and the upper perpendicular, and occupied throughout its entire length by a covered trough of sheet iron, one foot and a half broad, and six inches deep, passing down the middle of it. The lower end of this trough crosses the fire of the furnace, and is exposed to its full force; and the upper end terminates at the top of the chimney, in a hopper filled with the powder to be calcined. The flame and heated air in their passage up the chimney, impart their heat to this trough, which is closed on all sides; and the lower end, being in contact with the fire, is kept constantly of a full red heat. When sufficiently burned, the powder is removed by small portions from the lower end of the trough, by a door for the purpose; while the contents of the upper part descend and take its place. This door has apertures left in it to admit of a draft of air up the interior, and which acts upon the thin layer of powder lying on the floor of the trough. In another furnace of this kind, the chimney is bent like an elbow, not curved, the part next the fire being nearly horizontal, and the rest vertical; the first contains two troughs like that above described, one above the other, and the vertical part of the chimney is entirely filled with the material. Both these possess the advantage of heating the powder very gradually, by which the difficulties alluded to in the text are avoided.—Tr.

matter to reduce crude clay to fine powder; we know, that however dry it may be, this substance adheres to the bruising tools, and clogs underneath the stamper.

(XLII.) M. Bruyere, Inspector-general of bridges and roads, has bestowed much pains upon the conversion of clays into artificial pouzzolanas; his attention was principally turned to the method of giving these materials that degree of porosity, necessary for them to be quite equably acted on in all their parts by a moderate heat: he employed successively as ingredients for dividing them, lime, sand, and various vegetable substances. The results which he obtained, give us cause for serious regret, that he has not made them known by publication. M. Saint Leger has repeated most of M. Bruyere's experiments in the large way. He found, that a mixture of three parts of clay in powder, with one part of rich lime in paste, was that which we ought to give the preference to; and in this he is in perfect accordance with the intelligent Inspector-general. The artificial pouzzolanas which are formed by a slight calcination of clay mixed with lime, afford very active cements, which set in a few hours; but it is proper to add, that these same cements do not finally reach more than a moderate degree of hardness, they are besides very dear.

(XLIII.) We procured a very fine plastic clay, brought from the environs of Loupiac, (Lot,) and containing in 100 parts, 61 of silica, 31 of alumina, an inappreciable trace of the oxide of iron, and 8 of water. This clay, being pulverized and passed through a hair sieve, and equal portions of it put into hessian crucibles with covers let in, (and the joints carefully filled in with a luting of sand and refractory clay,) was in that state subjected to a good incandescence, for about half an hour, in the middle of a small dome furnace. The powders were always cooled in the closed vessels, and 100 parts gave by weight—

In the first experiment .	88.71	of which the mean is 88.543.
In a second ditto .	88.32	
In a third ditto . .	80.60	

Another portion of the same clay, equally well pulverised and sifted, having been calcined at a common red heat, on an incandescent metallic plate, and for 5 minutes, gave for 100 parts—

In a first experiment	89.85	of which the
second ditto	89.80	mean is 89.825.

A second portion, calcined in the same manner, but during 15 minutes, gave for 100 parts,

First experiment	88.50	of which the
Second ditto	88.65	mean is 88.575.

A third portion, calcined as before, but during 30 minutes, gave for 100 parts,

First experiment	88.45	the mean of
Second ditto	88.55	which is 88.50.

The very trifling difference which we observe between the weight of the clay when calcined in the close vessel, and that of the clay calcined in the open air, prove incontestibly that there is no kind of absorption. These differences may be accounted for; first, by an inequality in the duration and intensity of the heats; second, by the slight losses inseparable from a calcination in the open air, more especially when we are obliged to stir the substance. To what then is the real and notable difference in the binding qualities of the clay, as subjected to these two kinds of calcination, to be attributed? We should in vain endeavour to explain this.

At any rate, by calcination in a closed vessel, the clay is unable to acquire the faculty of yielding to acids the same quantity of alumina, as in the case of the ordinary calcination. The difference amounts to more than half. One part of the above plastic clay, being first calcined in a close vessel, and then, after eight days' exposure to the air, calcined again in an open crucible, lost $\frac{1}{100}$ of its weight.

One part of the same clay, being first calcined in the open air, and then treated in the same manner as the above, lost $\frac{8}{100}$ of its weight.

From this comparison it results incontestibly, that clays calcined in contact with the air, possess after cooling an absorbent property; which the same, when calcined in a close vessel do not.

But in what does this absorption consist? Is it simply hygrometric? This it is impossible for us to decide. The following, moreover, is a complete series of facts, observed in reference to cements prepared from two kinds of clay, calcined according to the two methods now in question.

	Time of setting.	Hardness, as measured by the fall of the needle, after		Hardness, as measured by the penetration of a piercer.	OBSERVATIONS.
		Five months' immersion.	One year of immersion.		
1.	Days.				
White clay, marked 2, in Table No. III., and calcined in powder in the open air .... 3.00 Rich lime in paste.... 2.00	1.50	3.25 mill. 0.1279 in.	2.50 mill. 0.0984 in.	3.40 mill. 0.1338 in.	The cements were tried by the needle at five or six millemetres (.196 in. or .236 in., Tr.) below the surface, and by the piercer, in the centre.
2.					
The same clay calcined in a close vessel....... 3.00 Lime the same 2.00	5	5.00 mill. 0.19685 in.	2.00 mill. 0.07874 in.	3.00 mill. 0.11811 in.	With a spring saw, they gave a nearly dry powder.
3.					
An ochreous clay, marked No. 1, in Table III., calcined in powder in the open air .... 3.00 Rich lime in paste .. 2.00	2.50	3.00 mill. 0.11811 in.	2.00 mill. 0.7874 in.	2.00 mill. 0.0787 in.	Numbers 1 and 3 were hard superficially, and did not at all adhere to the sides of the vessels containing them; Nos. 2 and 4 were deteriorated at the surface, to the depth of six millemetres (0.236 in., Tr.), and adhered powerfully to the sides of the vessels.
4.					
The same calcined in a close vessel .... 3.00 Rich lime as before.. 2.00	7.00	4.00 mill. 0.1575 in.	1.70 mill. 0.0669 in.	1.80 mill. 0.07086 in.	

(XLIV.) Is it to the presence of the potash, or the degree of calcination only, that the aquafortis clay owes its binding properties? In 1818, we subjected to the action of heat some small bricks, composed of clay moistened with solutions of the sub-carbonate of potash, containing from 1 to 15 per cent. of the salt, and could observe no relation between the energy of the products obtained, and the quantities of the alkaline oxides introduced; but on reflecting on it since, it has occurred to us, that by calcining the clay in the form of bricks, it was possible that, from the sudden action of the heat which that mode of calcination requires, the potash may have acted with too great intensity, and at last entangled the principles of the substance, by an incipient vitrification which, although imperceptible, might be no less sufficiently decided, to destroy all the binding property.

We in consequence recommenced our experiments, as follows. A hundred parts of a red effervescing clay, (containing silica 42.60, alumina 15.96, oxide of iron 8.00, carbonate of lime 24.64, water 8.80) were moistened in a solution of caustic soda, containing ten parts of saturated fluid; they were then dried, and brought to the state of powder, and afterwards calcined at a red heat in contact with the air. A hundred parts of slate, previously calcined in powder in contact with the air, and which in that state afforded only a pouzzolana very nearly inert, were moistened with the same quantity of the soda solution, then dried again, and calcined in the same manner as the clay above mentioned.

Thus prepared, the two pouzzolanas were mixed of a good consistency, with rich lime in paste, obtained by the ordinary extinction, and the products were immersed without delay. The phenomena which accompanied this immersion are exhibited in the following table.

COMPOSITION OF THE CEMENTS.	Time of Set.	Depression under the blow of the trial needle, after seven months' immersion.
	Days.	Mill.
Rich lime in paste, obtained by the ordinary mode of extinction 100 Pouzzolana, resulting from a simple calcination of the red clay in powder . . . 200	1.00	3.00 (0.118 in., Tr.)
Rich lime as before . . . 100 Pouzzolana, resulting from a calcination of the aforesaid clay impregnated with soda . . 200	1.00	1.75 (0.0689 in., Tr.)
Rich lime as above . . . 100 Pouzzolana, resulting from a simple calcination of the powdered slate 200	120.00	12.26 (0.482676 in., Tr.)
Rich lime as above . . . 100 Pouzzolana, resulting from a second calcination of the said slate, when impregnated with soda 200	6.00	8.0 (0.315 in., Tr.)

(XLV.)—* A patent has been recently taken out by Mr. R. F. Martin, and a company established for the manufacture of a newly discovered cement, prepared by calcining bricks composed of powdered gypsum, and the solution of the sulphate of potash (prepared by neutralising the alkali by the acid). The bricks are pulverised, after calcination at a red heat, and the cement thus formed is applied in the way of common stucco. It sets in about two hours, and some specimens which Mr. Martin had the kindness to exhibit to me, were of extreme hardness. As the material itself is of a pure white, it is admirably adapted as a vehicle for colours, and for the imitation of marbles and various descriptions of ornamental work, some of which are very beautiful. The cement is said to be quite uninjured by frost, and to be capable of enduring exposure to the weather without damage, for any length of time.—Tr.

NOTES ON CHAPTER IX.

(XLVI.) The reciprocal suitabilities of the ingredients of mortars and calcined cements were laid down in a general way, and for the first time, in the work which we published in 1818. The law which regulates these compatibilities has been since verified by a great number of Engineers, amongst whom we may name Colonel Raucourt, of Charleville. The treatise which he has printed, regarding the various kinds of lime employed in Russia, and on mortars in general, contains, besides, many interesting facts, of which we have availed ourselves.

We are now-a-days enabled to explain the contradictions exhibited by different memoirs relating to mortars and pouzzolanas; contradictions, which made the collection of facts known up to 1818 a real labyrinth, the clew of which had escaped the search of the most judicious investigations.

First example. The experiments made in 1786 at the harbour of Cette, by the commissioners of the States of Languedoc, on the occasion of Count Chaptal's researches into artificial pouzzolanas, established in an authentic manner, that the pouzzolanas of Vivarais were very inferior to those of Italy; and yet we find in the supplement to the first memoir on pouzzolanas, by Faujas de St. Fond, experiments also made with great care, and which demonstrate precisely the contrary. This is because, on the one hand, they used an eminently hydraulic lime, (the lime of Montelimart), and on the other, the rich lime of the environs of Cette. Now the eminently hydraulic lime, when used with the very energetic pouzzolanas of Italy, must, in fact, have afforded to Faujas, results inferior to those which he obtained at the same time from a feeble pouzzolana mixed with the same lime.

Second instance. The slaty schists, which were in 1807 advertised by G. Lepère, the Engineer, as substances capable of conversion by a powerful calcination, into pouzzolanas equal to those of Italy, deceived all the builders who wished to apply them to use. This was because the trial betons upon which that Engineer rested his comparisons, were made with the hydraulic lime of Grossvillé, and that lime, in fact, was adapted to a very slightly energetic pouzzolana, which ceased to be the case when rich lime was made use of.

Third instance. The military Engineers employed upon the works of Alexandria (in Piedmont), being ill satisfied with the induration of the beton prepared with the Casal lime and the powder of bricks once burnt, thought of calcining the powder violently in a reverberatory furnace, and they thus obtained better results than before; this was because, by giving a high degree of calcination to their artificial pouzzolana, they changed it into a pouzzolana of feeble energy, which in that state suited the lime of Casal very well, as it is hydraulic.

It would be tedious to enumerate in this place all the mistakes into which our predecessors have been led, by their uncertainty in regard to the mutual adaptation of the ingredients of calcareous mortars and cements. We must believe that Vitruvius, in making the qualities of lime to reside in its whiteness and expansion by slaking, merely viewed that substance in reference to its mixture with the excellent pouzzolanas of the environs of Rome, and not with pure quartzose or calcareous sands. Now, all who have written after him have not failed to repeat, in an absolute way, that the hardest and purest marbles furnish the best lime.

(XLVII.) A TABLE containing Twenty Compositions of Water Mortar, suited to different situations and circumstances. Extracted from Smeaton's essay on water cements. *Narrative of the Construction of the Eddystone Lighthouse.*

No.	WATER LIME WITH POUZZOLANA.	Lime powder.	Pouzzolana	Common sand.	No. of cubic feet.	Expense pr. cubic foot.
		Bushels.	Bushels.	Bushels.		s. d.
1	Eddystone mortar	2	2	—	2.32	3 8
2	Stone mortar	2	1	1	2.68	2 1½
3	——— 2nd sort..	2	1	2	3.57	1 7½
4	Face mortar	2	1	3	4.67	1 4
5	——— 2nd sort..	2	½	3	4.17	1 1
6	Backing mortar	2	¼	3	4.04	0 11
	WATER LIME WITH MINION.		Minion.			
7	Face mortar.........	2	2	1	3.22	1 5½
8	——— Calder composition	2	1	2	3.57	1 1
9	Backing mortar	2	½	3	4.17	0 10
10	——— 2nd sort..	2	¼	3	4.04	0 9½
	COMMON LIME WITH TARRAS.		Tarras.			
11	Tarras mortar........	2	1		1.67	4 0
12	——— increased..	2	1	1	2.50	2 9
13	——— further....	2	1	2	3.45	2 0½
14	——— still further	2	1	3	4.35	1 8
15	Tarras backing mortar.	2	½	3	3.50	1 2¼
16	——— 2nd sort..	2	¼	3	3.37	0 11¼
	COMMON LIME WITH MINION.		Minion.			
17	Ordinary face mortar..	2	2	2	4.75	1 5½
18	——— 2nd sort..	2	1	3	4.34	0 8¾
19	Ordinary backing mortar...............	2	½	3	4.05	0 8
20	——— 2nd sort...	2	¼	3	3.92	0 7½

The materials in the above mixtures were measured dry. The lime in powder thrown into the bushel and striked, but not beaten nor pressed down.

NOTES ON CHAPTER X.

(XLVIII.) We confined ourselves in Chapter I. to a simple definition, in what relates to "poor limes," and we have since forborne to speak of them for a very simple reason, viz., because they are very rarely employed, and because we are right in so doing: for, to the want of expansion of the hydraulic limes, they add all the negative qualities of rich lime. When, however, they are not absolutely devoid of hydraulic qualities, we may venture to use them, for want of others. In that case, it is evident that we ought to modify, in respect to them, what we have said regarding proportions; inasmuch as in an equal bulk, these limes actually contain a larger quantity of the earthy or metallic oxides, than the slightly hydraulic limes to which they are assimilated. Thus, M. de Laroche, Engineer of roads and bridges, in the experiments which he made at Brest, with a "poor lime" very slightly hydraulic, found, that the precept "it is better to err by defect than excess of lime" was not confirmed; because in fact, that precept is applicable only to the rich, and slightly hydraulic limes.

(XLIX.) This great question of the best method of slaking, has been for a long time agitated amongst builders. It is not astonishing that it has been decided by some in favour of immersion, and by others in favour of the ordinary process. Every one made use of the materials which fell to hand, without even suspecting, that the result which followed from any one such experiment, was applicable only with exactness, and rigorously, to these same materials. Faujas, alone, seems to have caught a glimpse of the fact, that lime, according to its nature, yields in preference to one kind of extinction rather than another; for he says, in a note of his memoir, speaking of immersion, "When we are in a situation to use an excellent quick lime," (we imagine that by quick lime, the author intended to denote an hydraulic lime,) " we may dispense with M. Lafaye's method; but whensoever we are compelled to use a lime of inferior quality, I strongly recommend its adoption."

We shall endeavour to explain, in what manner the management of the slaking after this or that process, exercises so great an influence on the qualities of cements of rich limes when immersed. It is certain, that all the principles which exist in quick lime, are still contained in it after slaking, since the smoke given out during the effervescence is, as we have shown, nothing more than vapourised water, and that we can therefore neither suppose a decomposition of any part of the water used in the slaking, nor the disengagement of the small quantity of carbonic acid, which the quick lime of commerce usually retains. Thus, the lime parts with nothing it contains. It merely absorbs water, and in addition, some carbonic acid, by the spontaneous extinction.

But it follows from the details given in Chapter V., that quick lime, when slaked spontaneously, or by immersion, is capable, at first, of forming a paste with much less water than what is necessary to fulfil the proper measure of its saturation. Now, as it retains the faculty of completing the dose by a prolonged immersion, it necessarily results, that when interposed in a mass of hydraulic cement, it must in a short time solidify all the water at liberty, and thus accelerate the set of the mixture, and at the same time increase its bulk. A very slight enlargement of size, the result of this action, very frequently exhibited itself in the course of our experiments, by the fracture of the vessels containing the immerged cement. The whole, therefore, explains itself by the aid of these considerations. This absorbent faculty of limes slaked spontaneously, or by immersion, ought evidently to be measured by the difference between the total weight of their proper allowance of water, and the weight of that which they have provisionally taken to form paste. Now this difference reaches its maximum with the very rich limes, and its minimum with the eminently hydraulic limes. Thus is explained, in the simplest manner, the progressive extension of the phenomena which we observe, and the correspondence between these phenomena and the different degrees of richness or poverty of the limes made use of.

(L.) But these considerations only embrace a certain compass of the scale, for they do not explain in what manner the ordinary mode of extinction rises superior to the two others, when we arrive at the hydraulic, and *à fortiori* the eminently hydraulic limes. It is true, indeed, that the difference between the first and second process is but trifling; but between the first and third, the difference is very remarkable. Now, viewing the hydraulic and eminently hydraulic limes as natural cements with excess of lime, we can conceive how the influence of an atmosphere, always more or less damp, must end by bringing on a chemical combination of the constituent principles [c], which combination would be the more energetic, the nearer these principles approach those which constitute the true natural cements. Now such is the case with hydraulic cements. Such limes then, when exposed for a long time to the air, end by becoming no more than a kind of *caput mortuum*, deprived of all binding quality.

(LI.) The strength of mortars, considered as aggregates, evidently resides in that of the hydrate of lime or matrix which envelopes the grains of sand. It is therefore very evident, that the greater the density of the hydrate is, the greater will be the resistance of the mixture, and that independently of all the molecular changes which may be caused by the

[c] An explanation different to that given by M. Vicat has suggested itself to me, and I shall venture to give it a place here, although I have not yet had an opportunity of ascertaining its propriety by any direct experiments. The energy of the hydraulic limes, as is well known, is developed by minute division, (note art. 185,) but entirely extinguished by the influence of a damp atmosphere. This I have imagined to be owing to a part of the lime uniting during calcination with the silica, alumina, magnesia, &c., and being thereby enabled to resist slaking for some time. When minutely subdivided, and made into a paste with water, these parts gradually combine with and solidify the interstitial fluid, causing the mixture to set and harden (note art. 307); and this takes place more perfectly, the more intimately the particles and fluid are brought in contact by the fineness of the subdivision. As it requires a

adherence which unites the lime and sand. For the result of these changes could only tend to increase the natural density of the gangue. Thus, without experiments, reasoning alone would have led us to the rule laid down.

(LII.) The cements cannot be assimilated with aggregates, (at least the cements properly made with pouzzolanas reduced to very fine powder). Here we have neither a sensible matrix, nor solid substances interposed at appreciable intervals. The mixture, whether it result or not from a chemical combination of the principles, is physically constituted as a homogeneous body. Therefore it signifies but little, before being mixed with the powdery ingredient intended for it, whether the lime be in a soft or firm paste, if, in the upshot, the mixture can be kneaded to a good clayey consistency; for, owing to their absorbing property, the pouzzolanas in general, almost always require an addition of water, and we then regulate the dose in such a manner as to attain the end proposed.

(LIII.) This method of extinction has been constantly applied at the works of the Angoulême bridge at Souillac, and the success of their foundations has completely justified its adoption.

(LIV.) M. Laguerenne, Engineer of bridges and roads, and in charge of the construction of the bridge of Charles 10th at Lyons, made a point of following, in every particular, the method of manufacture, and immersion, laid down in this chapter; he made use of a simply hydraulic lime, and of no

powerful effort to enable the water to overcome the cohesion of the semivitrified particles, the ordinary process is best adapted to the hydraulic limes, to which subdivision is so important. The process by immersion, or by exposure to the air, however, in which the action is less strong, merely causes those particles to *fall* or separate which are *un*altered by the heat, and consequently incapable of adding to the hydraulic quality of the lime; while the cementing matter gradually unites with water, but hardens and remains gritty, as the Sheppy stone does if wetted after calcination. Hence, the best possible plan of developing the virtues of hydraulic lime is, to slake it quickly, and then grind it to the finest powder, or (as is now frequently done) pulverise first, and then slake quickly.

other ingredient than sand and flints. His beton was immersed when cold, in a rapid current, and in an inclosure cleared out by dredging. And such was the success of this mode of procedure, that after 12 or 15 days, they were able to lay the foundations on masses of from two to three metres (6 ft. 6½ in. to 9 ft. 10 in., Tr.) in height. The enormous blocks of freestone of which the courses were composed, rested on the foundation of beton, as upon a rock.

The Lyonese builders, who frequently make use of beton in their foundations, slake the lime by aspersion, covering it with sand. They then pound and mix the materials rapidly with a large quantity of water, and employ the whole while still warm; this method succeeds upon the whole. When the beton is inclosed in the partitions of a framing, and when the foundation is mainly supported by the timber-work of such framing, it is of little consequence if the lime swells up, and absorbs water, and afterwards takes a long time to harden; the foundation does not stand a bit the worse; but it would have been altogether otherwise in the case which I have mentioned above. The slowness of the set, would hardly have allowed the foundation to be laid in less than a year or fifteen months after immersion; the rapidity of the current during this interval, would certainly have deteriorated the beton, and it is even probable, that its immersion might have been impossible; because the removal of the lime, would have left behind nothing but a residue of sand and flints, void of coherence.

(LV.) The works of the navigation of the Vezère, have just exhibited a sad example of the danger of using lime imperfectly slaked. Whether to gain time, or from any other motive, they thought that they might dispense with allowing the hydraulic lime they made use of, to stand and sour after slaking. The masonry exhibited nothing extraordinary as long as they remained dry during the low water in summer; but hardly had the winter floods set in, and submerged the works, than the mortar swelled, and with so much force, that the freestone blocks of the facing of the side walls were displaced in numbers, especially about the shoulders of the abutments. In short, the reconstruction of two locks was the result of these accidents.

(LVI.) The phenomena described in this paragraph are easily conceived; the water begins to exert its solvent action upon weak, or too rich cements, immediately after their immersion. It first acts upon the very thin superficial coating with which it is in contact, afterwards, it attacks the layer next below it, &c. But its action is retarded by the difficulties which the crust already attacked opposes to it, and as these difficulties increase in a rapid progression, the cement gains time to solidify more or less as its nature disposes it. Now since this solidification has a limit, and as the action of water, when continually renewed, is indefinite, the slow and incessant increase in depth of the soft crust of which we have spoken, necessarily ensues.

Mr. Petot, Engineer of bridges and roads, employed as an apprentice in 1825 on the works of the bridge of the Duke de Bordeaux at Saumur, has made some interesting experiments on the action of water at the surface of mortars of hydraulic lime and quartzose sand. He has observed, that the presence of siliceous matter in the state of quartz, exerted, during the earlier period of immersion, a powerful influence, which would seem to indicate a molecular action which it would be highly important to establish. The following is a concise table of Mr. Petot's experiments.

Nature of the compounds experimented upon.	Quantities of lime dissolved, in 1000 parts of the water of immersion, after three days.
1. Hydrate of hydraulic lime slaked spontaneously . .	1.030
2. Mortar, composed of 100 of sand, and 100 of the said lime	0.560
3. Mortar, composed of 150 of sand, and 100 of the hydrate	0.540
4. Mortar, composed of 200 of sand, and 100 of the hydrate	0.607
5. Mortar, composed of 250 sand, and 100 hydrate . .	0.601
1. Hydrate of hydraulic lime, slaked by the ordinary process	1.100
2. Mortar, composed of 100 parts of sand, and 100 of the said lime	0.840
3. Mortar as before, composed of 150 sand, and 100 of the hydrate	0.500
4. Mortar as before, composed of 200 sand, and 100 of the hydrate	0.450
5. Mortar as before, composed of 250 sand, and 100 of the hydrate	0.630

" It is evident from this table," says Mr. Petot, " that it is not by diminishing the number of points of contact, that we diminish the solubility of the lime; because in that case, it would be necessary that that solubility should decrease, in proportion as the quantity of sand is augmented, which is contrary to the facts."

The portions of the mother water which were regained for each sample, were about 130 grammes (2007.7 grs., Tr.) sometimes more, and sometimes less than the half of the whole with which each of them was covered; they had been filtered, and precipitated by the binoxalate of potash. It might be objected, that the same specimen, although quite homogeneous, would, if divided into many perfectly equal parts, and covered with different doses of water, give as many different solubilities; for each portion should yield the same weight of lime from an equal surface of contact, or nearly so; but this remark ceases to have any foundation, when the differences between the quantities of water, and, more particularly, those quantities themselves, are trifling, as in the case before us.

The relative solubilities of the different specimens being in perfect accordance with the apparent hardness acquired by each specimen at the time of trial, it thence follows, that the proportions of sand corresponding to the minimum solubility, are also those which confer upon the mixture the greatest hardness.

(LVII.) The observations contained in this paragraph, evidently apply merely to the first ten or fifteen years following immersion. We are unable to estimate the influence of ages; it creates between the constituent elements, even of mixtures of rich lime and quartz, placed in certain circumstances, reactions, which it is difficult to explain: reactions from which in time results a cohesive force, which leaves nothing to be desired. But as we have said elsewhere, mortars which continue weak for more than one century, are the same to us as if they would never harden; for our bridges, locks, piers, dams, &c., have to resist floods, thaws, and bad weather, sometimes even before their entire completion.

NOTES ON CHAPTER XI.

(LVIII.) We made choice of a calcareous sand, of similar grain to the granitic sand of the Dordogne, for the purpose of preparing with it two mortars of the following proportions:—

No. 1.	Hydraulic lime in paste . . .	100.00
	Calcareous sand	100.00
No. 2.	Lime as before	100.00
	Sand do.	150.00

We also prepared in the same way, two other mortars in the like proportions, marked below No. 1 *bis*, and No. 2 *bis*, with granitic sand. These various mixtures being divided into bricks, and some exposed to the weather, others to the action of a damp soil for fourteen months, gave the following absolute resistances per square centimetre:—

		Kil.	Corresponding resistances, per English square inch.—Tr.
Mortars exposed to the air . .	No. 1 .	15.19	216.21 lbs. avoir.
	No. 1 *bis*	12.24	171.22 ”
Ditto	No. 2 .	16.99	241.83 ”
	No. 2 *bis*	16.80	239.13 ”
Mortars buried in the ground . .	No. 1 .	12.72	181.05 ”
	No. 1 *bis*	12.00	170.80 ”
Ditto	No. 2 .	13.68	194.72 ”
	No. 2 *bis*	12.48	177.64 ”

Thus the superiority is always on the side of the calcareous sand; though at the same time we observe that the differences are trifling.

(LIX.) Whether it were the effect of chance, or that the Romans were acquainted with the mutual relations between the qualities of the lime, and the size of the sand; they frequently made use of unequal grained sand, with feebly hydraulic mortars, and very coarse sand, sometimes even small gravel, with rich lime. The remains of aqueducts, amphitheatres, baths, &c., which we find at Cahors, Vienne, and other places, prove this. Monge, in visiting the ruins of Cesarea in Syria, found sunken impressions of mouldings and ornaments in the heaps of mortar belonging to the counterforts which sustained the remains of a temple dedicated to Augustus; the reliefs were chipped at the edges, but the mortar projected. It was so hard, that Monge in vain tried to break off a bit. Now this mortar was composed of very fine sand and a small quantity of lime, which by its greyish colour, we may presume to be hydraulic. The old ramparts of Viviers, (Ardéche,) which they were unable to destroy except by mining, were cemented with a mortar composed of very fine sand mixed with a white, eminently hydraulic lime. This lime, of which we still continue to make use at the present day, is generally known along the Rhone by its excellent quality. It is probable, that in these cases chance has led to the choice of the sands, more than intention; but this does not prevent these very examples, *as facts*, lending their support to our own experiments, and confirming the rules laid down in Chapter II.

(LX.) Loamy or argillaceous particles, deprive powders and sands, which are largely impregnated with them, of the faculty of composing mortars capable of enduring exposure to the weather, in conjunction with the powerfully hydraulic limes, and, à fortiori, with the rich limes. The old canal of Nivernais, is a remarkable instance of this. The locks and other works of art built about thirty-six years ago, with mortar of slightly hydraulic lime and arenaceous sand, were, when the works of that canal were recommenced, in a deplorable condition. It is especially opposite to the pond at Baye, that the bad quality of these mortars is most remarkable. Nought is to be seen amidst the rubbish of the dilapidated walls, but a reddish powder, in which we in vain endeavour to find a solid fragment as big as a nut.

(LXI.) The explanation of the effects of slaking upon mortars of rich lime, when exposed to the air, appears to be as simple as the case of mortars immersed. We see, in fact, that the mortar which has received the smallest allowance of water, is also that which has the least to lose by drying, and in a word that there is more substance, and consequently more density in any bulk of mortar, the lime of which has been slaked by immersion or spontaneously, than in an equal bulk of mortar, the lime of which has been carried to the farthest possible limit of expansion by the ordinary process of extinction. In support of what has been said on this subject in App. XXII., we shall subjoin the following facts. 100 parts by weight of rich lime slaked by the ordinary process, after fifteen months' exposure (in small quantity) to the air, weighed 59.60. A hundred parts of the same lime, and in paste of equal consistency, which had been prepared by immersion, weighed at the same period 77.10. Now the two hydrates had at this time arrived at that point, at which all external influence had ceased; that is to say, that their weight varied only in an insensible degree, by a purely hygrometric action, sometimes positive, and sometimes negative; it ought here to be remarked, that the bulks were very nearly equal at the commencement of the experiment. Thus the density of the hydrate of lime slaked by the first process, would finally, if shrinking were impossible, have been no more than 0.77 of the density of the same lime slaked by immersion. The superiority of the spontaneous over the ordinary extinction is explained and understood in a precisely similar way; but the differences of density cannot be referred to, in comparing the extinction by immersion with the spontaneous method; the superiority of this latter mode rests upon causes which are at present unknown.

With the simply hydraulic and eminently hydraulic limes, the cohesive property surmounts all the rest; and the differences of density, at all events very small, are in a measure lost, in comparison with the adhesive action arising from the perfect division of the material. Moreover, the negative influence of the spontaneous extinction has no other cause, than that which has been specified in App. L.

(LXII.) Opinions are pretty nearly unanimous, as to the bad qualities of a mortar which has been drowned. The ancients insisted on the necessity of working up the mortar without any addition of water; good mortar, said they, ought to be tempered only with the sweat of the mason[d]. It is the fact, that all the power, and all the activity of the cleverest labourer, will not succeed in uniting a very dry sand with a very stiff hydrate of hydraulic lime (these two substances being in the proportions comprised within ordinary limits);

[d] General Treussart gives an opposite opinion; and as it would, if correct, contribute materially to the economy of the manufacture of mortars, it is desirable that it should not be entirely neglected without being put to the test. He says, "The following experiments will show, that those are in error who pretend, that mortar should be made with the sweat of the workmen only. It is sufficient that the sand be well mixed with the lime, and this mixture can be made more effectually, and much more economically, when the mortar is in a rather thin paste, than when it is stiff. Besides, there is no inconvenience in making it rather thin, since it frequently becomes stiffer than what it is required to be when the masons use it; because, as I have remarked above, quick-lime when reduced to thin paste, retains for a tolerably long period the faculty of

in such a case it is necessary to add water, and there will be no harm whatever in doing so, provided we do not exceed the proper quantity. Good preparation of mortar, is an excessively hard thing to obtain in a work-yard: this difficulty has suggested to different builders the idea of making use of machines, and thereby rendering the manipulation independent of the will or strength of the workmen. Amongst the more or less ingenious attempts which we might notice, that which seems to have best succeeded, consists in making a strong wheel, built upon the common pattern, to roll over the lime and sand placed in a circular trough. The movement is given by one or two horses harnessed to a gin. This method, designed by M. Saint Leger, has been employed for the fabrication of hydraulic mortars for the canal of Saint Martin; it is right to add, that in the mortar so worked, the lime and sand are very exactly mixed, and always of the same consistency; that the economy of workmanship is considerable, but it is also certain, that the mixture is far from possessing that degree of stiffness which it ought to have; moreover, we can never attain it by the millstone, the wheel, or the harrow. It is to the stamper alone that we must resort to arrive at a solution of this problem, and its solution is not without difficulties. In fact, it is not sufficient to beat with force and rapidity, it must be struck true, that is to say, in such a way that the blows be not lost. Now it is this which constitutes address, a quality of which machines are not susceptible, except up to a certain point.

The bad quality of drowned mortar is not merely owing to its want of compactness; it depends also upon a chemical cause, which it is important to be acquainted with, viz., that

solidifying water." (Mémoire sur les Mortiers Hydrauliques, p. 13.) It ought not to be forgotten, however, that most of General Treussart's experiments were made with mixtures of common lime, and tarras, or pouzzolana. Now the drowning of the mortar would in such a case only deprive the lime of its power of absorbing water, of which property the set of a compound of rich lime and pouzzolana is quite independent. The case may, however, be very different with hydraulic limes, if they owe any part of their virtues to the faculty of solidifying water.—Tr.

the water which is employed in large quantity, tends to decompose the silicate of lime with excess of base, and reduce it to the state of the neutral silicate. M. Berthier has remarked, in fact, that if we take hydraulic lime as it leaves the kiln, and composed for instance of 555 parts of pure lime, to 400 of silica, if we slake and agitate it with a large quantity of water, we shall only be able to collect by filtration 615 parts of undissolved matter, or neutral silicate, composed in that case of 400 parts of silica, and 215 parts of lime [e].

In using the water in small quantity, on the other hand, it would be absorbed and solidified in great part by the combination, so that decomposition would not take place. Now there can be no doubt but that the hydrosilicate of lime, in which the whole mass is chemically united, is of more use than a mere mixture of the neutral hydrosilicate and the pure hydrate of lime, which tends to form when too great a quantity of water is made use of.

Nevertheless, it may not be impossible to communicate to a mortar, mixed thin, the compactness which results from the stiff consistency given at first; to effect this, it would be best to spread it out in the sun, or in the open air, for the purpose of evaporating the superabundant water quickly; and afterwards to beat it with stampers, in a trench properly disposed. It will be curious to learn how far a mortar thus prepared, may be equivalent to one obtained directly by a good manipulation. This problem is peculiarly interesting in reference to the economy of the labour of its preparation.

(LXIII.) This method of guarding the mason's hand against the causticity of the lime, was discovered by Mr. Silguy, Engineer-in-chief of bridges and roads, employed upon the canal from Nantes to Brest; the workmen found it answer so well, and valued its efficacy so highly, that they would at

[e] The silicate of lime is a compound of lime and silica; the silicate with excess of base, is a compound in which the lime exceeds its atomic combining proportion. The neutral, that in which they neutralise one another.—Tr.

last have gone on to an abuse of it, if care had not been taken.

The soaking of the materials is evidently an addition to the labour, which should be taken into account in the detailed estimates. It is true it is sufficient to water the compact stones, such as granite, quartz, mill-stone grit, lime-stone, marble, &c., at the moment of using them; but a mere aspersion will not answer in regard to spongy and absorbent substances, such as bricks, the soft or arenaceous lime-stones, sand-stones, &c.; when we have such materials, it is necessary to moisten them without ceasing, and to keep them in a permanent state of imbibition. To effect this, and to avoid imposition, it is best to water in mass the heap from which they are taken, so that they may reach the mason's hand in a soaked state. In large work-yards, a fire engine, which projects and spreads the water to a great distance, answers perfectly for this purpose. It was thus that Mr. Inspector-General Deschamps, at the bridge of Bordeaux, watered the bricks piled up on the service bridges.

(LXIV.) The influence of slow desiccation upon the goodness of mortars of hydraulic limes, has been for a long time known in Italy. At Alexandria, in Piedmont, they manufacture artificial stones which they call prisms, because being generally used for the construction of the angles of walls, and the starlings of bridges, &c., they have, in fact, the form of a triangular prism. For this they make use of an hydraulic lime, quarried in the neighbourhood of Casal; they slake it after the ordinary mode, and when it has soured five or six days, they put it into the middle of a basin of irregular grained sand, from the substance of common sand, to that of coarse gravel. This sand is chiefly quartzose, containing calcareous débris; the mixture is now made, and upon this they bestow considerable pains. Before using it, they prepare a triangular prismatic trench of an arbitrary length, on a level space, secure from floods. They smooth the sides with the trowel and a little water, and form the prisms in it in successive layers, inserting in the mortar stones of uniform size regularly distributed. The prisms are then covered with the same earth dug out of the trench, so as always to give a thickness at top of 30 centimetres. (11.8 in., Tr.) The proportions for a cubic metre (35.3 cubic ft., Tr.) are 0.24 c. (8.47 cubic ft., Tr.) of lime in paste, 0 m. 90 c. (31.77 cubic ft., Tr.) of uneven grained sand, and 0 m. 20c. (7.06 cubic ft., Tr.) of pebbles.

They give the prisms a length of 1.40 c. (4 ft. 6 in., Tr.) to a breadth of side of 0 m. 80c. (2 ft. 7½ in., Tr.); they remain under ground usually for three years, but two are sufficient when the lime is of the best quality; after that time they are removed and applied to use. They will then bear heavy loads; we have seen them thrown one on the other, from the height of six or seven metres (19½ to 23 ft., Tr.), they were chipped at the edges, but did not break.

We left a piece of mortar composed of common granitic sand, and common lime slaked spontaneously, but coated with an hydraulic cement, to prevent the immediate contact of the fluid, under water for the space of a year; after which it was withdrawn, deprived of its coating, and placed upon the damp floor of a cellar, and afterwards, little by little, removed to more elevated situations. After some months, the exterior parts of this brick appeared very hard; we transferred it suddenly from the cellar to a loft, in order to hasten the period of its desiccation a little, and some time afterwards it was subjected to experiment. At the moment of its rupture, the part detached divided into two portions, one of which separated from the other, just as the yolk of an egg separates from the white, when it is well boiled. The envelope was pretty hard, but the interior was readily squashed in the fingers. We again exposed the piece in which this separation had not taken place, to the action of the air, expecting certainly, that at the end of a few months the nucleus and the crust would exhibit no difference, but it was not so; the nucleus never attained the hardness of the envelope. The difference was, and always continued to be such, that there was a very marked interruption of continuity between the two parts. Thus the act of solidification had been rudely interrupted, by the transition from a fresh damp atmosphere to a warm one. This fact was too interesting to allow us to trust to one experiment. Such as we have since repeated have all exhibited the same phenomena.

Vitruvius (lib. ii. chap. viii.) gives many instances of buildings near Rome, ruined in a short time, owing to the too hasty drying of the mortar. The lime, says he, separates from the sand, if the stones (of the masonry) absorb all the moisture by their pores.

(LXV.) M. John (in the memoir quoted) mentions, that not long ago, on demolishing one of the columns of the tower of Saint Peter, at Berlin, built about 80 years, and 27 feet in diameter, they found the mortar in the interior of the masonry as fresh as if it had only been applied the day before; it had the caustic taste, and formed milk of lime with water. But we very often meet with good mortars in the foundations, and in the massive walls of the buildings of the middle ages. We have often discovered the nature of the lime made use of for these mortars, by the particles or lumps of the lime not mixed with the sand; and we have generally found it to be either rich, or very feebly hydraulic. It results from this, that after a maceration of six or seven hundred years, favoured by the constant humidity of the soil under which they are buried, the mortars of rich lime at last harden: the term is rather long. With reference to these observations, we shall add the analyses, made by M. John, of some old and antique mortars of excellent quality.

TABLE OF THE CONSTITUENT PARTS,

CONTAINED IN 100 PARTS OF MORTAR.

DESCRIPTION.	Carbonic Acid.	Lime.	Silica in solution.	Water.	Grains of quartz, with 2 or 3 per cent. of sand alloyed with clay and oxide of iron, separated by washing.	Alumina and iron dissolved.
1.—Mortar six hundred years old, from the cathedral of Brandebourg	5.00	8.70	1.25	1.30	83.75	
2.—Mortar six hundred years old, from the church of Saint Peter at Berlin, (the foundation laid in a situation constantly impregnated with water,)	1.75	9.25	3.75	6.75	78.50	
3.—Roman mortar, from an ancient city wall, built at Cologne under Agrippa, in the 1st century of the Christian era . .	9.00	15.16	0.25	4.00	68.00	2.75
4.—Roman mortar, from an ancient tower built by Agrippa, at the same period	12.00	24.00	0.25	5.00	56.00	2.75
5.—Roman mortar, taken from a beton built in the Rhine . .	2.25	6.90	0.35	1.00	89.50	

By rendering these results, approximately, into technical language, we find as follows[f]:

[f] Being anxious to compare the quantities of lime and sand exhibited by the analyses of old mortars, with the proportions used for the public works, I made some experiments, in order to determine the ratio between the weight of *pure* lime, and sand, and the measures of hydrate of lime, and sand, in the condition in which they were usually served out for the mixture of mortar, for the Government buildings. I found, that a cubic foot of slaked lime in that condition, (vide note to Appendix 20,) and weighing 30 lbs., contained only 18½ lbs. of *pure* lime; an equal measure of sand weighed 75 lbs., which numbers are very nearly in the

The composition of mortar No. 1 . .	Hydraulic lime in paste .	100
	Quartzose sand	560
That of No. 2 . .	Eminently hydraulic lime in paste	100
	Quartzose sand	604
Ditto No. 3 . . .	Rich lime in paste . . .	100
	Quartzose sand	187
Ditto No. 4 . . .	Very rich lime in paste .	100
	Quartzose sand	137
Ditto No. 5 . . .	Feebly hydraulic lime . .	100
	Quartzose sand	607

This reduction is on the supposition—

1st. That in all probability the silica, alumina, and iron dissolved, belonged to the lime.

2d. That the cubic metre of quartzose sand weighs 1400 kil. (3088.8lbs., Tr.), and that the same bulk of quick lime without voids, also comes to 1400 kil., (3088.8 lbs., Tr.).

3d. That the simply hydraulic lime, gives 1.500 for 1.00 by slaking; the eminently hydraulic lime 1.00 for 1.00, and the rich or feebly hydraulic lime 2.00 for 1.00.

These hypotheses being in accordance with the average results of experiment, must approach very near the truth.

This table evidently proves, that the influence of the qualities of the lime, and the proportions, is lost in the presence of that of ages, for all the five mortars analyzed, were very hard, especially Nos. 3, 4, and 5. But this astonishing effect of time, is only felt, as I have before remarked, by mortars in foundations, or such as are lodged in the centre of massive

proportion of 1 to 4. Hence, if the quantity of *pure lime* furnished by an analysis, be multiplied by 4, its weight will then be to the weight of silica and insoluble matter of that kind contained in the cement, in the same ratio as the *bulks* of hydrate of lime and sand would be, if measured in the way I have described in the note above referred to. This rule is however only applicable to analyses of mortars of rich lime, as the varying ratios of expansion in slaking, of the poor and hydraulic limes, cause the same experiments to be necessary, for the elimination of the like rule, for their reduction to technical language.—Tr.

bodies of masonry; in a word, where the humidity can have been constantly retained.

NOTES ON CHAPTER XIII.

(LXVI.) Our observations have led us to think, that the cements of artificial pouzzolanas, obtained by calcination of the clays wholly or nearly free from iron, are those whose deterioration is the most marked, when we pass them suddenly, or even by degrees, from a damp situation to a dry atmosphere; the oxide of iron, therefore, is favourable to cements exposed to the air. The observations which we have made on this kind of phenomenon, are too few to give much weight to this opinion, which it may nevertheless be of use to express.

(LXVII.)—* "Three parts of oil heated with one sixth of its weight of litharge, and one part of wax, form a good composition for protecting valuable work. One part of linseed oil, one tenth of its weight of litharge, and from two to three parts of resin, form a suitable composition for common work.

"The oils are heated, and applied with a brush over the surface to be preserved; should the surfaces be impregnated with humidity, they must be heated and perfectly dried before painting them. I have made great use of the boiled lithargirated oils, applied hot, on cornices and parts exposed to moisture, and I have in all climates met with unexpected success; the stuccoes and the walls thus defended, were dry and sound, and resisted both rainy weather and frosts. With respect to compositions containing tar, bitumen, resin, wax, &c., &c., they ought always to be heated before application."

"All the pigments composed of oils and the ordinary colours, and which have little body, require to be renewed after a few years; but of those containing wax, resin, &c., a single coat is sufficient; and if the time should arrive when they cease to be of any effect, the hydraulic mortars will by that period have acquired so much solidity, as to oppose sufficient resistance to the usual agents which tend to deteriorate them."—Raucourt de Charleville, Traité des Mortiers, pp. 283, 4, and 293.

(LXVIII.) In Italy even, where the climate is remarkably mild, the vertical plasters of pouzzolana cement exposed to the north, finally become ruined[g]; they repair them by means of rich mastics variously composed. The weak cements overcharged with lime, or those resulting from a bad manipulation, crumble, or become disaggregated by efflorescence, in the same way as ill-burnt bricks. Cements of good quality split in flakes. These effects of the frost are established by cases unfortunately too authentic. If, in fact, we consult the military Engineers, as to the pointing in cement applied to the revetments of the forts in the north of France; the Engineers of roads and bridges, regarding the same ce-

[g] During my superintendence of public buildings in the northern division of the Madras Presidency, I made many endeavours to investigate the cause of the very speedy decay in certain situations, of stuccoes, which in others, apparently similarly exposed, possessed the greatest durability; and I was soon led to observe the very injurious effects of alternations of moisture and dryness upon mortars composed of rich lime and sand, by noticing the almost constant failure of those mortars which exhibited a *mottled* appearance on first drying. This phenomenon, which is interesting when viewed in connexion with the future durability of a cement, consists in a distinct exhibition of the courses of the masonry through the plaster, the joints being clearly defined, sometimes on a darker, and at others on a lighter ground, and indicating at one time an excess of humidity in the spaces opposite the bricks, and at others opposite the joints. These appearances were generally confined to within a few feet of the lower portion of the walls, which parts alone were subject to the early decay I allude to; and as they take place in a climate not subject to frost, and denote a cause of the degradation of mortars hitherto unexamined, it may be useful to notice briefly the leading facts which were observed.

Common stucco, when plastered on a wall of brick and mortar, generally presented the appearance, on first drying, of damp joints and dry spaces (opposite the bricks); and this seemed to be owing to the spongy nature of the bricks, which, unless fully saturated with water in using them, caused the stucco to part more speedily with its moisture in those parts over them, than elsewhere. After the same stucco has once thoroughly dried, if the situation be damp, these appearances may be reversed, more especially if the bricks be of an absorbent kind, or ill burnt. In this case it appears that being kept constantly damp by sucking up moisture

ments applied to locks; and all the proprietors of manufactories, or others, who have been in the way of using them, whether in vertical plastering, or in terracing, &c., &c.,—all will answer in the affirmative.

But the intermixture of sand powerfully checks the expansive force of the freezing water, by assimilating the cements, whose texture it modifies, to porous and permeable stones; to the sand-stones, sandy lime-stones, &c. In fact, it is not the most compact and most impermeable stones, that are the least acted on by the frost; experience always proves the contrary.

The action of frost, presents a class of phenomena which appear incomprehensible, when we look upon them as the mere results of the expansive force of the ice. With the

from the earth, they become more charged with it than the stucco itself, which is of a less permeable nature; and exhaling it during hot weather, render the *spaces* damp in comparison with the joints, which have remained dry. And it is remarkable also that, in such a case, the phenomena may be again reversed by soaking the whole surface with water; for if the stucco be not a very compact one, so as completely to exclude imbibition, the subsequent desiccation will proceed so much more rapidly opposite the bricks, which will then be comparatively *dry*, that the joints will continue long damp after the spaces have ceased to be so. Stucco, covering walls built of brick and *clay*, I have always seen exhibiting dry spaces and damp joints.

Roman cement, Mr. Loriot's composition stucco, and such as *set* on first applying them, I have generally observed to present an opposite appearance to common mortars in which the lime has been well tempered. The best mortars, when exposed in situations which occasioned the above phenomena, were liable to early ruin, sometimes within a week of their completion, the decay being invariably confined to the region in which they were manifested; such as were beyond the reach of the moisture remaining perfectly sound for years. The failure was accompanied by the separation of large flakes, which were detached from the body of the cement, leaving the mass in a crumbling and disaggregated state behind it. These flakes varied in thickness with the age of the mortar, a fact which led me to the supposition, that they might consist of the superficial crust of the regenerated mortar, separated from the part behind it by some internal changes, occasioned by the alternations of moisture and dryness; but I found it impossible to prove this point directly, from the difficulty of detecting the separation of a patch of the falling plaster, in time to find the parts in contact with it still in a caustic state.—Tr.

exception of such stones as are intersected by threads, visible or not, but into which the water may insinuate itself in thin capillary seams, we have hitherto been unable to give any explanation of it. In fact, as often as we endeavour to approach this question, the most contradictory facts present themselves at the same time. Thus, wretched mortars of rich lime, very much overcharged with sand, and which a mere pressure of the finger would crumble, stand a cold of 12° centigrades[h], (about + 10° Fahrenheit, Tr.,) with impunity, although soaked to saturation; whilst the brick, twenty times harder, falls to powder at 4° or 6° (from 25° to 21° Fahrenheit, Tr.). Some lime-stones of a compact kind ("vives"), very hard, intersected in every direction by short closely crowded veins, (as for instance, the Chouin lime-stone, used at Lyons,) endure the most severe winters, at the same time that other stones physically alike, (such as those of Saint George's at Cahors,) split in every direction. Lastly, what can we say of those materials, which are liable to injury from frost when impregnated with the water they contained in quarry, and not liable to be acted on after they have lost it, although they may be elsewhere impregnated to saturation by rain water?

In this state of the case, we are compelled, in order to speak with certainty, as to the qualities of such and such materials, to note the manner in which they behave for several years. And even many years may be insufficient in the southern countries, where severe winters ensue only at long intervals. We must therefore try to multiply the causes of destruction, to the specimens which we make trial of, and gain time by accumulating, as it were, the effects of many winters. To effect this, we shall observe, that after having raised or separated the parts of the substances it acts upon, the crystallized water, for all that, keeps them still clinging to one another, until a thaw; at which period the disunion takes

[h] I take this of course, to refer to *minus* 12° of the centigrade thermometer; and I have given its equivalent accordingly, as about + 10° Fahrenheit, or 22° below freezing point. Were it *plus* 12° instead of minus, it would correspond to very nearly 54° of our thermometer.—Tr.

place. Hence we infer, that we ought to take the duration of the frosts much less into account, than the number of successions of frosts and thaws which follow one another. Now we can create as many thaws as we have days of frost in a winter, by each time pouring boiling water over the experimental specimens, to melt the ice or snowy efflorescence with which they are covered, and release them from their detached flakes, if they have any.

It was in this way that we proceeded with respect to a number of fragments of mortars and cements of all kinds, which we studied. The results, which are made known in the thirteenth chapter, are the fruit of the observations of ten consecutive winters, amongst the number of which, is that of 1819 and 20, when the thermometer fell to 12° of the centigrade, (about 10° Fahrenheit, Tr.)

(LXIX.) Whilst we were engaged with these researches, Mr. Brard, (then director of the collieries of Lardin, in the department of Dordogne,) tried to distinguish the stones which are injured by frost, from those which are not so, by substituting for the expansive force of congealing water, that of an easily crystallizable salt, the sulphate of soda. No sooner were we made acquainted with this happy idea, than we were eager to try it upon our mortars. In conformity with the directions of the able mineralogist whom we have just named, a hundred specimens were impregnated with a warm saturated solution of the sulphate of soda, and then exposed in a loft to the temperature of 25° to 30° centigrade, (77° to 86° Fahrenheit, Tr.,) and lastly, washed from day to day with pure water. These specimens were not long in giving signs of degradation, in the following order:—

AFTER TWENTY-FOUR HOURS.

All the mortars of rich limes have become slightly disaggregated. The progress is more perceptible, and more widely diffused, in the set of *ordinary* (as regards the slaking) specimens, than the *immersion* (as to extinction) series. The *spontaneous* (extinction) set are the least injured.

The mortars of hydraulic lime and fine gravel, or very coarse sand, split at the corners; those with fine sand, stand well.

AFTER FORTY-EIGHT HOURS.

The three series of mortars of rich lime, are in decomposition, with the exception of the numbers of the *spontaneous* series, which correspond to the proportions of 50, 60, 70, and 80 parts of sand, to 100 of lime in paste. The other "rich" mortars detach thick flakes and split at the corners; the "poor" ones become disaggregated superficially.

The mortars of hydraulic lime, and gravel, or very coarse sand, have completely crumbled to pieces; those in which the sand is fine, continue to stand well.

AFTER SEVENTY-TWO HOURS.

None of the mortars of rich lime remain, except numbers 1 and 2 of the *spontaneous* series. Of the mortars composed of hydraulic limes and fine sand, the "richest" split at the corners, the "poorest" continue to do well.

The observations were continued for 24 days; after the twelfth, not one of the mortars remained unimpaired, except those of very poor hydraulic limes, and the mortar of the *spontaneous* series of rich lime, corresponding to the proportions of 50 sand to 100 of lime in paste, which continued well till the twenty-third day. We have thus been convinced, that the agency of the alkaline salt is much more powerful than that of water frozen in our climate. We were about to have instituted other experiments made with solutions variously charged, when an accident diverted us from it. In trying, in fact, to recover from the washings the portions of salt dissolved by the successsive eleutriations, we perceived that the form of the crystals was no longer the same. Instead of six-sided prisms, we got nothing but crystals flattened into buttons, and entirely of a different structure. No doubt then, the alkaline ley had acted upon the lime of the mortar.

M. Brard's process therefore can no longer be applied to this description of materials. We should, however, recommend it, when we only wish to establish the relative qualities of cements or mortars. It will then perfectly indicate the order in which the compounds will be able to resist the action of cold; and if we moderate the effects of the crystallization, by impregnating the bodies under trial with a solution not saturated, it is probable that some day we may succeed in discovering that degree of solution which suits bricks and stones, to represent with precision, in respect to them, the power of congealed water, at this or that degree of the thermometer; but it is hardly any where, except in the north, that such an experiment can well be made.

We have said, that mortar composed of rich lime slaked spontaneously, and compounded in the proportions of 50 sand to 100 of lime in paste, continued to do well until the twenty-third day; and this at a time when excellent mortars of hydraulic lime, and calcareous stones of acknowledged goodness, did not last beyond the seventh or eighth day; such an experiment deserved to be repeated, and it was so, with the same success. The mortar which behaved in this manner was six years old, its relative resistance by superficial *centimetre* corresponded to 4.80 kil. (68.32 lbs. avoirdupois per square inch, Tr.) This is not, as we may see, quite half of the mean relative resistance of mortars of hydraulic limes [i]. Now there are a multitude of cases in reference to ordinary buildings, in which the mortars of revetments have absolutely nothing to contend against but the weather, and are besides sufficiently strong, when there is nought to fear from the frost.

We do not think, that in this respect it is possible to devise any thing in the way of a mixture of lime and sand, which offers a better assurance than the mortar above mentioned.

[i] Vide Article 282.—Tr.

NOTES ON CHAPTER XIV.

(LXX.) Loriot's process consists in introducing into mortar, worked to a thin consistency, such a quantity of quick lime in powder, that the superabundant water of the mixture may suffice for its extinction, and that consequently, the mixture, without reaching that degree of desiccation producing pulverulence, may nevertheless become solid in a few instants.

Loriot's process, like many others, has been very much in vogue, and like many others also, has ended by falling into oblivion. Such is the usual fate of all false conceptions, and that was certainly one, which viewed the induration of mortars as the mere result of a more or less rapid desiccation, and in consequence, supposed it possible to obtain this end, by the introduction of a powerful absorbent[k].

Specimens cast in mortar of hydraulic lime prepared according to M. Loriot's process, although afterwards submitted to the influence of a very slow desiccation under a fresh soil, never attained more than a moderate hardness. It seemed, that the introduction of the quick lime in powder, however it might have hastened the set, had not made up for the want of density inseparable from the thin consistency which

[k] Mr. Smeaton says of this process, "I have made trial of this method, both in small and in large; for however little likelihood of advantage a proposition may contain, yet, when this concerns a physical process, nothing can be safely concluded but from actual trial: and I must candidly own that the effect was much better than I had expected; for I found the composition not only set more readily than mortar as commonly made up, but much less liable to crack, and consequently, if this cement was made use of in water building, it was less apt to re-dissolve, because it would more speedily get set to a firmer consistence, and so as more ably to resist the water from entering its pores; but when the water was brought upon it, in whatever state of hardness it was at the time, it at best remained in that state without any further induration, while the water remained upon it; and as I expect would so remain, till it had some opportunity of acquiring hardness by further drying."—*Construction of the Eddystone Lighthouse.*

In operating with the pure shell-lime of Madras, I have myself found we were obliged to give the mixture of sand and slaked lime in the first instance.

NOTES ON CHAPTER XV.

(LXXI.) " In 1796, Messrs. Parker and Wyatt obtained the royal patent for the manufacture in London of a particular kind of lime, which they termed *water cement*, and to which they afterwards gave the name of *Roman cement*, a name the more unsuitable, as the Romans neither were acquainted with, nor ever made use of any thing of the kind. This speculation has had the greatest success, and it has given birth to several others of the same kind, which prosper equally.

" M. Lesage, a military Engineer, made known, twenty years ago, the properties of a kind of lime which they made use of at that time at Boulogne-sur-Mer, (Dover straits,) and which he has designated by the name of plâtre cement. We learn from the very circumstantial report of it which he published in the " Journal des Mines," vol. XII., p. 145, that that lime is precisely the same material as the English

this process answer very well for the composition of stucco; for after a comparison during two years, of surfaces similarly exposed, containing from 120 to 150 square feet each, I found M. Loriot's composition very superior in hardness, and in resistance to atmospheric changes, to the ordinary stucco of *well tempered rich lime* and sand. It is, however, very difficult to get workmen unaccustomed to it to apply it properly, as a little want of care mars the whole effect. It is also more expensive.

The mortars above alluded to were worked up and applied with jaghery (coarse sugar) dissolved in water. Mr. Smeaton agrees with Dr. Higgins in thinking, that both for dry and water works the use of *fresh* slaked lime, of whatever kind, is preferable to M. Loriot's process; and I am quite of this opinion myself, with respect to *hydraulic* lime quenched hot from the kiln and pulverised, which will be found to produce as good an effect with less labour and uncertainty.—Tr.

natural cement. M. Drappiez has given a very exact analysis of the Boulogne stone, which we shall now compare with that furnished by an analysis of the English one.

		English stone.	Boulogne stone.
" Carbonate of lime		657	616
Carbonate of magnesia		5	..
Carbonate of iron		60	60
Carbonate of manganese		19	..
Clay,	Silica	180	150
	Alumina	66	48
	Oxide of iron	..	30
Water		13	66
		1000	970

" The English stone is of a brown grey colour, compact, fine grained, and susceptible of polish; its specific gravity is 2.59. That of Boulogne is also compact, very fine grained, and susceptible of polish, but it is of a yellowish grey colour; it has never been met with except in rolled pebbles on the sea beach; while the English stone is dug out from the marls, where it is found imbedded in the form of nodular masses.

" Saint Petersburg has now, like London, its natural cement; it owes this advantage to Messrs. Clapeyron and Lamé, French mining Engineers, temporarily attached as professors to the Polytechnic Institute of Russia. This discovery has already effected important saving in the execution of the great hydraulic works to which the natural cement has been applied." (Extract from *M. Berthier's Memoir on hydraulic lime-stones.*)

We have ourselves met with natural cements on the banks of the Loire, between Nevers and Briare, and at Baye, at the point of junction of the canal of Nivernais; but in their quickness of setting, and final induration, they were not equal, either to the English cement, or that of Boulogne.

The following is an analysis of the cement of Baye.

Bluish grey Argillaceous Lime-stone.

Carbonate of lime	320
Silica } Alumina }	526
Carbonate of magnesia	40
Alumina and iron, dissolved	94
Water	30
	1010

This cement set three days after immersion, and at the end of one year exhibited a hardness, measured by the depression of 3.55 mill. (0.13976 inch, Tr.) [1]

NOTES ON CHAPTER XVI.

(LXXII.)—* Extracts from the analysis (by Dr. Malcolmson) of the cement from the pyramid of Cheops.

(1.) A hundred grains of this cement, having been pulve-

[1] The following is an analysis of the cement referred to in the note to Art. 24, and the mineral composing which was brought from Bezoarah in the Guntoor district of the Madras presidency. It is found on the banks of the river Kistnah, and in the immediate vicinity, as I was informed, in the form of small irregular shaped nodules, of a dark bluish grey colour, of various shades, very hard, susceptible of a dull polish, specific gravity 2.52. The constituents in 100 parts were, lime 43.5, silica and alumina 18, magnesia 2, carbonic acid 36. The silica and alumina which remained behind on dissolving the stone in acid, presented the appearance of a black clay; but the lime was of a buff colour after calcination.

This cement, when made into stiff paste, and immersed in the manner prescribed in the beginning of this volume, (Articles 8, 9, 10, 11, 12,) set in 48 hours; and in 15 days was so firm, that I was unable to thrust the needle (Art. 20) into it, and was obliged to desist for fear of breaking it. At the end of *nine months*, its hardness was measured by the instrument used by M. Vicat, and was indicated by a depression of 0.15 of an inch; no part of the surface having been removed previous to this trial.—Tr.

rised, and dried at a steam heat, were immersed in six ounces of pure water, and after standing over night, were heated in contact with it, for the purpose of removing its soluble salts. It was then separated from the water by filtration, washed, and dried, and on weighing amounted to 81.5 grains, having lost 18.5 grains of soluble matters, which were examined separately, and found to consist of sulphate of lime 15.3, sulphate of soda 3.2.

(2.) The residue of 81.5 grains, left after the separation of the above soluble salts, was now treated with dilute muriatic acid, (of which 4 cubic inches were used,) and lost of carbonic acid 4.7 grains. If to this be added 2 cubic inches of the gas, (or 0.94 grains,) for the absorption by the 4 in. of fluid, the whole weight of carbonic acid in the cement will be 5.64 grains[m].

(3.) The insoluble parts which were left behind after the action of the muriatic acid in the last process, were now separated from the solution by filtering, and after being washed and dried, were found to weigh 54.7 grains. This consisted principally of crystals of sulphate of lime, with alumina[n].

(4.) The solution just mentioned, and washings, &c., having been tested for magnesia, iron, silica, and other matters, unsuccessfully, was now boiled with excess of carbonate of potash, which threw down a precipitate, (of carbonate of lime, &c.,) which when washed, dried, and weighed amounted to 19.9 grains; and was reduced by ignition to 11 grains. Of

[m] Particular attention was paid in this analysis, to the determination of the exact quantities of carbonic acid and lime, and the processes were repeated, and varied once or twice, with close agreement in the results. Part of the cement was also, previous to analysis, reduced to powder, and tested by moistened turmeric paper, the solution of sulphate of iron, and the tincture of galls, &c., to detect the presence of caustic lime, but no trace was observable.

[n] The insoluble residue mentioned in process 3, was levigated with oil and powdered charcoal, and the whole then ignited. On digesting part of the powder which remained after this process, first in caustic potash to separate alumina, and then (after washing) in dilute nitric acid, the whole was dissolved without any residue; whence I concluded that it consisted of lime and alumina only.—Tr.

this, by a subsequent process, 0.3 is found to be alumina; the remainder pure lime.

(5.) In order to ascertain the amount of water in a hundred grains of the cement, that quantity, after being dried at 212°, was put into a bent glass tube, one portion of which contained chloride of calcium, and which was connected at its extremity with the mercurial trough. On applying heat, much water was given off, a very small quantity of which (estimated at about one grain) escaped from a leak in the apparatus. After completing the process, the chloride was found to have gained 16.5 grains in weight, to which if we add the one grain which escaped, we shall have the whole quantity of water separated from the cement, which will amount to 17.5 grains. This, however, includes both the water associated with the carbonate of lime, and that contained in the sulphates of lime and soda.

The constituents of 100 grains of the cement are as follows:

Soluble salts, consisting of sulphates of lime and soda	18.5 grains.
Carbonic acid	5.64
Lime	10.7
Alumina	0.3
Insoluble substances, consisting of alumina and crystals of selenite	54.7
Water associated with the carbonate of lime, and loss	10.16
Total grains	100.00

(LXXIII.) There have been discovered at Euriage, not far from Grenoble, the remains of an ancient fish-pond, the borders of which were thickly incrusted [o]. When submitted

"In May 18—, I fixed a stick on a mass of travertine covered with water, and I examined it in the beginning of the April following, for the purpose of determining the nature of the depositions. The water was lower at this time, yet I had some difficulty, by means of a sharp pointed hammer, in breaking the mass which adhered to the bottom of the stick;

to examination, the incrusting matter dissolved completely in nitric acid, the ferro-prussiate of potash occasioned a slight precipitate of the Prussian blue, and the muriate of barytes also precipitated a small portion of the sulphate of barytes; whence it results, that independently of the lime, the fragment experimented on contained also a little iron and sulphate of lime.

It is certain that the ancient cements, separated from incrustations, would not possess sufficient consistency to be cut into thin plates, and afterwards take a polish. The thickness of the coating of carbonate of lime which covers these cements is very variable, in some parts of the channel of the aqueduct of Gard, it amounts to 0.05 m. (1.968 inches, Tr.)

(LXXIV.) The massive masonry of the ancient Bastile of Paris could only be destroyed by mining; a few years ago, there were still to be seen at Agen, near the Gravier-gate, the remains of a bridge which was thought to be old from the hardness of its mortar: they were obliged also to use powder to remove the remains of a pier which was in the way of the promenade. This bridge, the construction of which the lovers of the marvellous would gladly have referred back to the Pelasgi, was built in 1189, in virtue of a charter of Richard the First, king of England, and at that time ruler of a part of France. The mortar of the bridge of Valentré, built at Cahors in 1400, is in every respect similar, both as to the quality of the lime, the proportions, and the size of the sand, to that of an ancient theatre, the ruins of which are to be found in the same town, six or seven hundred paces from the river. Frequently repeated trials have shown only a very slight difference between the resistance of antique and old mortar; and that difference is entirely in favour of the latter.

it was several inches in thickness. The upper part was a mixture of light tufa and the leaves of confervæ; below this was a darker and more solid travertine, containing black, and decomposed masses of confervæ; in the inferior part, the travertine was more solid, and of a grey colour."—Sir H. Davy, Consolations in Travel, p. 127.—Tr.

NOTES ON CHAPTER XVII.

(LXXV.)—* A detail of the facts alluded to, relative to the induration of magnesia, will be found in Note 2 to Appendix V., and the following are the particulars of a similar experiment with the powder of sulphate of lime (plaster of Paris).

One hundred and thirty-six grains of the common calcined sulphate of lime were diffused in water, and then put aside and allowed to set and harden. A few days afterwards, the solid cake was removed, and exposed for several hours to a steam heat, after which it appeared quite dry. It was now weighed, and found to have increased 18 grains, to which an additional allowance of about 2 grains was due, on account of the quantity dissolved in the water of immersion. The cake was afterwards reduced to powder, and again exposed to the steam bath for about an hour, without perceptible change, and it was then re-immersed, but found to be incapable of again solidifying under water.

As this experiment was undertaken merely with the view to ascertain the fact of the solidification of water during the set of the sulphate of lime, no pains were taken to measure the ratio of combination, in order to which, it would have been necessary to have used the sulphate in the precise condition of being entirely freed from water, but at the same time not over burned. Mr. Graham (Transactions of the Royal Society of Edinburgh, Vol. XIII. p. 313) states, that when calcined at 270°, it is entirely anhydrated, and is then capable of recombining with two equivalents of water, which would amount to 36 grains to 136 of the sulphate. When calcined at too high a heat, it refuses entirely to recombine with it, and is technically termed *burnt stucco*.—Tr.

(LXXVI.)—* The following particulars of the experiments alluded to, are extracted from my memoranda.

Experiment 1.—Two parts of Rajahmundry clay, (a white

kind of pipe-clay, which did not effervesce, or lose weight, by diffusion in nitric acid,) were calcined at a red heat, on an iron plate, for half an hour, and then made into a stiff paste with slaked rich lime, and set in six hours.

2.—Three hundred grains of the *same* calcined powder were immersed at the same time, and kept under water for twelve hours. The water was then poured off, and the powder dried at a steam heat, and weighed; when it was found to amount to 299.5 grains, the 0.5 gr. having been lost in the process. Hence it is evident that, during the period of solidification of the above cement, there was no absorption of fluid to which to attribute it.

3.—And the following experiment proves, that the *compound* also does not combine with, and solidify water during its set; (unless indeed, it be explained by supposing a union so feeble as to be decomposed at a steam heat;) for 300 grains of a similar calcined clay, and 300 grains of well slaked rich lime, after being made into a stiff paste, and immersed, set in six hours, but were retained in the water for two days; after which, on being taken out, and dried at a temperature of 212° F., the weight was found to be 599 grains, a loss of one grain having occurred in making the powders into paste and mixing them.

4.—A quantity of the Rajahmundry clay of Experiment 1. calcined in powder, as in that experiment, was kept in water for a month and eight days, and then taken out and made up into a stiff paste with half its weight of slaked rich lime; and it was found on immersion to set in six hours as before.

5.—A mixture in stiff paste, composed of two parts of a calcined clay powder, and one part of slaked rich lime, and which set in three hours, was allowed to remain immersed for six weeks. It was then taken out, reduced to fine powder, sifted, and again made into a stiff paste and re-immersed. It set the second time very slowly and imperfectly, but was strong enough to bear the needle in six days. On first solidification it soiled the finger when touched, and though possessing some cohesion as a mass, appeared to have but little intimate connection between its particles, as the surface was easily destroyed by gently rubbing it. These defects, however, were not permanent; they ceased to be conspicuous after the first month of immersion.—Tr.

(LXXVII.) This work had been completed, when M. Girard de Caudemberg, Engineer of roads, published, under the unassuming title of "Notice," a lengthened memoir on hydraulic mortars composed with arenaceous sands. The positive results made known in this memoir having been communicated to us a long time since by the Author, are contained in the preceding general compendium. We shall therefore confine ourselves in what follows, to noticing some novel assertions, which appear to us to be in manifest contradiction to well established facts; and further, to the discussion of the degree of probability of the theoretical inductions, by means of which M. Girard has attempted to explain the solidification of mortars and cements.

M. Girard says, in relation to cements of *calcined* arènes, compared with cements of arènes in the natural state, "that after one year there was no appreciable difference between the consistency of one and the other, excepting in the external parts, to the depth of one or two centimetres (.39 or .78 inch, Tr.); that, consequently, the only advantage afforded by calcination, is that of hastening, in a remarkable manner, the period when the set of the betons is completed.

The fact may be true, but the consequence which M. Girard deduces therefrom cannot be general, as it is founded only upon one particular case of calcination. This Engineer, in fact, confined himself (page 25 of his Memoir) to heating his arènes till they changed colour, in a sheet iron *evaporating* basin; while he ought to have pushed the heat to a strong incandescence, and kept the substance in that condition for fifteen or twenty minutes; M. Girard would then have obtained a *highly energetic* artificial pouzzolana, instead of a *simply energetic* one, which his process afforded.

We pass over in silence some minor details, the object of which is to compare the price of the cubic metre of mortar of hydraulic lime and common sand, with that of the cubic metre of rich lime and calcined arène; because an error or exaggeration in a matter of that kind is not likely to be dangerous, experience being always prompt in doing justice.

In the second part of his Memoir, M. Girard developes Macquer's system, and attributes the excess of the resistance of mortar over that of its gangue, to the necessity the aggregate is placed in, to break with a jagged surface in lieu of a plain and even one, which would be that which the gangue would assume in its fracture, were it not mixed with sand.

We have, in Chapter XVII., seen what objections we have made to this hypothesis, which is not even supported by the first and indispensable experiments, whereby we ought to endeavour to ascertain what the ordinary absolute resistance of a prismatic solid, taken in the direction of its axis, will be, when owing to any obstacles whatever, the plane of its fracture is obliged to assume various degrees of inclination towards that axis.

We may make up for this deficiency, though imperfectly, by saying, that from 0° to 45° the resistance remains sensibly proportional to the area of surface. (This assertion is derived from numerous experiments made upon various conditions of the rupture of solids of inextensible fibres, of which experiments we submitted an account to the Academy of Sciences, at its sitting of the 11th December, 1826.) This being admitted, it is easy to satisfy ourselves with a lens, or the naked eye, that the limit of 45° is far from being attained by the average inclination of the various faces exhibited by the sections of fracture of the mortars of fine sand; and that in consequence, the development of the linear outline taken from these sections, can never be in the ratio of 1414 to 1000[p], to that which would be exhibited by a matrix em-

[p] 1.414, &c., is the square root of 2, which is the ratio of the hypothenuse of a right angled *isosceles* triangle to its base; and consequently also, the ratio of the development of any number of these lines forming an irregular jagged section, to the base they stand on, when their constant inclination to it is 45°.—Tr.

ployed without sand, and supposed to be *smooth and plane* in its fracture. Now, not only does the hydrate of hydraulic lime not show a smooth fracture, but the inequalities it presents seem to be pretty nearly of the same kind as those of the mortars of fine sand. After this, how can we account for the proportion of 3.6 to 1.0 afforded by experiment, between the resistance of a good mortar of fine sand, and that of its matrix.

We shall then fail to explain the solidification of good mortars, as long as we persist in denying to sand a very notable influence upon the cohesion acquired by the hydrate of hydraulic lime which envelopes it. This influence is besides placed beyond a doubt by Mr. Petot's experiments (App. LVI.).

In the latter part of his memoir, M. Girard discusses several facts relative to hydraulic cements composed with rich lime and clay, calcined, or in the natural state; and arrives at this conclusion, "that the solidification of mortars of argillaceous pouzzolana, depends upon the combination which takes place, between the lime and the silica on the one hand, and between the lime, the alumina, and the oxide of iron on the other."

We shall refrain from following M. Girard through this discussion, 1st, Because it does not comprehend all the known facts respecting artificial pouzzolanas; and amongst these facts there are some very important ones which are far from accommodating themselves to the hypothesis laid down. 2dly, Because we consider the method adopted by the author, for separating the silica from the alumina and oxide of iron in ochreous clays, to be inadequate; and that there is reason to suppose that a combination of silica and alumina has been constantly taken for silica alone, a circumstance which brings to nought all the consequences drawn from so inexact a supposition[q].

[q] M. Girard says (p. 57), that he separated the various clays which he examined, into silica on the one hand, and alumina and oxide of iron on the other, by means of hydrochloric (muriatic, Tr.) acid, and ammonia.

We agree with M. Girard, that it is impossible to disallow a chemical action in the solidification of cements; but we also think, that the question which has for its object the determination of how, and between what principles, that combination particularly takes place, is still to be solved. These observations, and those which precede them, do not amount to more than the researches of mere theory, and in no way invalidate the important and positive facts made known by M. Girard regarding the hydraulic properties of the arènes. The title which this engineer has acquired to the esteem and gratitude of builders, is therefore not the less deserved, nor less secure.

Now clays treated by hydrochloric acid in excess, do not, even after many days' digestion, part with more than a portion of their alumina. Some even, scarcely yield 2 or 3 per cent. of it.—*Original Note.*

THE END.

TABLES.

TABLES.

ACCOUNT OF THE MANNER IN WHICH THE EXPERIMENTS WERE MADE.

All the compounds, whether mortars or cements, (unless when the tables expressly mention the contrary,) were kneaded stiff and, as nearly as possible, to the same degree of consistency, by the aid of a pestle.

The compounds intended for immersion were put into rather deep than broad cups, sometimes of glass, sometimes of Delft, or common glazed earthenware, and were covered with pure water immediately after their preparation.

The compounds intended to harden underground, or in the air, were either buried, or exposed under sheds, in the form of quadrangular prisms, having for section a rectangle of four or five centimetres (1.575 in. to 1.97 in., Tr.) base; to 25 to 40 millemetres (0.98 in. to 1.97 in., Tr.) of height.

The quickness of set of the immerged compounds was measured by the number of days which elapsed, from the instant of immersion, to the moment when the surface of the substance was able to bear, without any appreciable depression, a knitting needle of 0.0012m. (0.47 in., Tr.) in diameter, filed at right angles at one of its extremities, and loaded at the other with a leaden knob, of the weight of 0.30 kil. (10 oz. 9 drs. avoir., Tr.)

The relative hardnesses of the same compounds were generally ascertained by the penetration* of a steel pin, very slightly tapered, and terminated at its lower extremity, perpendicularly to its axis,

* We imagined at one time, that we could deduce from a certain number of experiments given in our first work, "that the squares of the numbers expressing the depths of penetration of a needle impelled by a shock into any substance, are reciprocally proportional to the relative or absolute resistances of

by a small plane circular surface of 0.166 cent. (0.0653 in., Tr.) diameter. This pin, loaded with a weight of 0.9961 kil. (2 lbs. 3 oz. 2½ dr. avoirdupois, nearly, Tr.), fell freely from a constant height of five centimetres (1.97 in., Tr.).

Sometimes also, we made use of a piercer similar to that which Perronet employed in measuring the hardness of stones ; the indications set down in the tables in that case make express mention of this.

The abolute resistances of the prisms which were buried, or exposed to the air, have been deduced from the relative resistances of these same prisms by determining from a sufficient number of experiments, and for the dimensions laid down, the actual value assumed by the co-efficient represented by $\frac{1}{5}$ in the formula of Galileo.

Some Engineers are in the habit of delaying the immersion of the trial betons, until they have allowed them to acquire a firm consistency in the air. The results thus obtained are altogether delusive, and cannot be compared with what takes place on a large scale.

The experiments which have for their object the comparison of the quickness of set of the various betons, require great care, and are liable to causes of anomaly which we have it not always in our power to avoid. Whatsoever efforts we may make to give the same degree of consistency to the paste of the betons which we wish to compare, it is extremely difficult at all times to obtain any great degree of exactness, because, according to the process of slaking, and the nature of the lime, and the ingredients, we are obliged to make use of different quantities of water. Now it takes very little, either in excess, or defect, to retard or accelerate the set by a day or two.

We have not, moreover, been able to make all our experiments in the same season ; so that some betons were immerged in the violent heat of summer, and others at a mean temperature ; a circumstance which must have differently influenced the quickness of set. We have, in fact, had occasion to remark, that hydraulic

that substance;" and according to that principle, the depths of penetration were transformed, by calculation, into numbers proportional to the resistances ; but after the very judicious observations of M. Vauthier, the Engineer, we have resolved upon returning to the numbers expressing the depths.

mortars and cements, immersed in warm water at 40 degrees, (104° Fahrenheit, Tr.,) were capable of supporting the trial needle in a few hours, while corresponding ones kept in a flowing stream at 7 degrees, ($44\frac{1}{2}$° Fahrenheit, Tr.,) could not do so till after many days. But these variable influences only sensibly affect the period of first set, and when that period was required to be an essential element of comparison in our researches, we took means to avoid all the causes of error.

The measure of the relative resistances was also attended with numerous difficulties, which depended principally upon the smallness of the prisms, and the unequal influence of the carbonic acid of the atmosphere upon their different faces, according to their particular position during the period of consolidation.

It was, besides, necessary to take into account the difference in the dimensions of these prisms, not only because in each solid the sum of the particles regenerated by carbonic acid is not proportional to the whole bulk of the substance, but also because the relative resistances are not, as was till now supposed, proportioned to the relation between the square of the height of the section of fracture, to the length of the prism under experiment.

We shall conclude by remarking, that when we compare the resistances of mortars which fulfil various functions in a building, we ought to have regard to the change of place and condition which we cause them to undergo in submitting them to trials. Mortars, and some stony substances which are able to absorb a certain quantity of water, lose in such case, from $\frac{1}{5}$ to $\frac{2}{5}$ of their strength; whence it follows, that mortars buried underground, or those which we take from the foundations of an edifice, to compare with others, ought to be tried at once, if we wish to ascertain the real resistance which they were capable of in the very spot they occupied.

The machines which were made use of in measuring the resistances of mortars and cements, are represented, with explanatory notes, in plates 2 and 3 placed after the tables.

TABLE No. I.

IN SUPPORT OF CHAPTER I.

Comparison of the Qualities of various Limes, with the Chemical Composition of the Lime-stones furnishing them.

DESCRIPTION OF THE INGREDIENTS.	CONSTITUENTS IN 100 PARTS. Carbonate of lime.	Alumina.	Silica, impalpable.	Silica, in the condition of sand.	Carbonate of magnesia.	Oxide of iron.	Oxide of manganese.	QUALITIES. Time of set, in days.	Resistance measured by the penetration of a steel pin into the surface of the hydrate of lime.	REMARKS.
LIME-STONES FURNISHING RICH LIME.										
Crystallized carbonate of lime.	100	....	....	...	....	....	....	indef.	none.	All the hydrates submitted to experiment, had a year's immersion. Most of the analyses are by myself, the rest are due to Mr. Berthier, Engineer in chief. The written statements of the experiments, from which most of the annexed results have been extracted, are certified by the engineers attached to the canals of the Isle, Rance, and Loire.
Carrara marble	100	....	....	....	....	....	....	do.	do.	
Incrustations on the walls of the aqueduct of Gard.......	99.15	0.30	0.45	....	....	....	....	do.	do.	
Fresh water lime-stone of Chateau Landon, compact, yellowish colour, slightly vesicular	97.00	0.40	0.60	....	2.00	....	....	do.	do.	
Lime-stone from Cartravers (Côtes du Nord), lamellar, of a blueish grey colour	92.42	1.00	3.67	....	2.91	..	....	do.	do.	
LIME-STONES FURNISHING POOR LIME.										
Sandy lime-stone of Calviac (Dordogne)	70.00	1.25	2.00	24.75	....	....	....	indef.	do.	
Coarse lime-stone of Dessin (Loire Inferieure)	61.89	1.57	3.10	26.00	7.44	....	....	do.	do.	
Lamellar lime-stone of Villefranche (Aveyron), colour ochreous.................	60.09	....	....	....	30.30	*3.00	†6.00	....	......	* The iron is in the state of carbonate. † The manganese is in the state of carbonate.
LIME-STONES FURNISHING FEEBLY HYDRAULIC LIME.										
Coarse compact lime-stone of Quilly (Loire Inferieure) ..	74.60	3.10	4.90	....	17.40	....	....	days. 14.00	inches. 0.3307	The needle penetrates the hydrates by depressing their substance, without producing splinters.
Tufa of Vitry, near the banks of the Loire (Nièvre).......	90.00	3.40	6.20	....	....	0.40	....	9.00	0.4397	
Shelly lime-stone, from the neighbourhood of Songy (Nièvre)..................	88.00	4.00	6.85	....	....	1.15	....	15.00	0.5472	
LIME-STONES FURNISHING HYDRAULIC LIME.										
Secondary lime-stone from Nismes (Gard), compact, yellowish	82.50	5.00	8.40	....	4.10	....	....	4.00	0.2126	
Ochreous yellow lime-stone, of Champ Vert (Nièvre)	82.63	6.00	11.00	....	....	0.37	....	3.00	0.1988	
Coarse lime-stone from Pompeau (Isle et Vilaine), called "brule-mort vert"	75.83	4.00	9.50	....	9.00	1.67	....	4.00	0.2291	
Marly lime-stone, from the hillock at Poids de fer, on the banks of the Loire (Nièvre).......................	82.00	7.00	8.96	....	....	2.04	....	6.00	0.2382	
Marly lime-stone of Baraigue (Lot), bituminous, blueish grey colour	82.25	5.50	10.50	....	....	1.71	....	5.00	0.2205	
LIME-STONES FURNISHING EMINENTLY HYDRAULIC LIME.										
Castina (flux) from the plains of Barrée, near Sardy (Yonne)..................	65.00	8.00	26.25	....	....	0.75	....	3.00	0.1181	The hydrates of eminently hydraulic lime, all splinter under the blow of the proving needle.
Marly lime-stone of Devay, near Décise (Nièvre).......	80.50	7.20	12.00	....	....	1.30	....	4.00	0.0736	
Marly lime-stone from the valley of Riégeo (Nièvre)......	78.00	6.00	14.00	....	....	2.00	....	3.00	0.1181	
Semi-compact lime-stone of Fertotot, near Bec d'Allier.	80.75	6.50	11.50	....	....	1.25	....	8.00	0.0945	
Yellow lime-stone, of the domain of Fleury, near Décise.	77.40	8.35	13.25	....	....	1.00	....	3.00	0.1551	

TABLE No. II.

IN SUPPORT OF CHAPTER V.

Comparison of the Hardness and Absolute Resistance of Compounds, resulting from the Union of Water with various Limes.

NAMES.		Absolute resistances in pounds Avoirdupois per English square inch. In the condition of hydrate, fresh dried.	Absolute resistances … At the expiration of one year.	Relative hardness of the compounds. In the condition of hydrate, fresh dried.	Relative hardness … At the expiration of one year.	Specific gravities of the compounds. In the condition of hydrate, fresh dried.	Specific gravities … At the expiration of one year.	Of the lime-stones which furnished the lime. Specific gravity.	Of the lime-stones … Relative hardness.
		lbs.	lbs.						
EMINENTLY HYDRAULIC LIME. FROM BOUGARDE (TARN ET GARRONE).	Slaked by the first process.	15·4	66·0	0·034	0·054	1·172	1·212	2·060	0·096
	Ditto by the second ditto.	5·98	59·5	0·029	0·049	1·202	1·252		
	Ditto by the third ditto.	1·85	44·5	0·024	0·025	1·042	1·153		
THE SAME. FROM MONTELIMART (DROME).	Slaked by the first process.	39·3	96·6	0·067	0·100	1·342	1·432	2·367	0·363
	Ditto by the second ditto.	14·2	75·1	0·064	0·092	1·253	1·348		
	Ditto by the third ditto.	13·1	54·0	0·042	0·050	1.100	1·127		
HYDRAULIC LIME. FROM THE ENVIRONS OF NISMES (GARD).	Slaked by the first process.	34·1	89·4	0·044	0·110	1·304	1·462	2·500	1·000
	Ditto by the second ditto.	17·1	67·6	0·041	0·082	1·236	1·253		
	Ditto by the third ditto.	11·4	41·9	0·037	0·070	1·104	1·122		
THE SAME. FROM SAINT CERÈ (LOT).	Slaked by the first process.	27·3	85·1	0·048	0·080	1·193	1·227	2·653	0·777
	Ditto by the second ditto.	15·6	51·2	0·040	0·075	1·157	1·183		
	Ditto by the third ditto.	4·55	28·5	0·036	0·037	1·063	1·170		
SLIGHTLY HYDRAULIC LIME. FROM CABESSUT (LOT).	Slaked by the first process.	30·7	89·5	0·051	0·090	1·183	1·222	2·573	0·700
	Ditto by the second ditto.	12·8	44·9	0·049	0·082	1·233	1·275		
	Ditto by the third ditto.	10·7	31·3	0·040	0·060	1·096	1·212		
COMMON RICH LIME.	Slaked by the first process.	41·3	97·5	0·121	0·200	1·554	1·656	2·653	1·000
	Ditto by the second ditto.	11·7	37·6	0·033	0·050	1·114	1·145		
	Ditto by the third ditto.	27·8	51·2	0·039	0·050	1·029	1·043		
VERY RICH LIME.	Slaked by the first process.	92·2	129·8	0·222	0·500	1·611	2·008	2·462	0·154
	Ditto by the second ditto.	13·6	92·5	0·060	0·170	1·007	1·354		
	Ditto by the third ditto.	61·5	126·4	0·080	0·250	1·385	1·410		
EMINENTLY RICH LIME.	Slaked by the first process.	125·9	127·2	0·166	0·500	1·493	2·621	2·532	0·666
	Ditto by the second ditto.	27·3	92·2	0·121	0·160	1·136	1·500		
	Ditto by the third ditto.	102·5	126·7	0·133	0·200	1·572	1·654		

The hardnesses were measured by the action of a piercer; the numbers by which they are expressed, are reciprocally proportional to the depths of the holes, made by a certain number of revolutions under a constant pressure.

TABLE No. III'.*

IN SUPPORT OF CHAPTER VII.

Comparison of the Action of Muriatic Acid upon Clays taken in different conditions.

DESCRIPTION OF THE CLAYS.	Constituent principles of the Clays, in 100 parts.					Portions of these principles dissolved by digestion in Muriatic Acid, in 100 parts of the Clay, taken.								
						In the natural state.			Calcined in powder, exposed to the air.			Calcined in powder, in a close vessel.		
	Silica.	Alumina.	Carbonate of lime.	Oxide of iron.	Water.	Alumina.	Oxide of iron.	Carbonate of lime.	Alumina.	Oxide of iron.	Carbonate of lime.	Alumina.	Oxide of iron.	Carbonate of lime.
White plastic clay of Loupiac (Lot) (2)	61.00	31.00	..	..	8.00	2.85	..	..	12.40	..	..	5.48	..	..
Grey effervescing clay of Souillac (Lot) (3)	43.63	32.90	2.16	9.29	12.02	6.00	6.46	2.13	18.20	6.60	2.12	not made.		
Brown ochreous clay, found infiltrating the fissures of calcareous rocks (1)......................	42.06	28.67	..	12.33	16.94	6.04	5.35	..	17.40	7.00	..	7.20	4.15	..
Ochreous clay, of a blood colour, separated from an arene (4)	43.40	26.00	..	18.60	12.00	2.30	13.86	..	15.00	9.80	..	not made.		
Red clay, colour of wine dregs, from Baye (Nièvre) (5)............	42.60	15.96	24.64	8.00	8.80	1.75	3.51	24.64	2.00	3.00	24·60	not made.		

* Table No. III. will be found on the page following.—Tr.

TABLE

IN SUPPORT OF

COMPARISON OF THE QUALITIES OF HYDRAULIC CEMENTS, with

DESCRIPTION OF THE SUBSTANCES.	Constituent principles in 100 parts of the Substances in the natural state.						
	Silica.	Alumina.	Oxide of Iron.	Carbonate of Lime.	Lime.	Potash and Soda.	Water.
VERY ENERGETIC SUBSTANCES.							
Tarras from Andernack, containing	46.60	20.60	12.00	..	3.00	5.00	12.80
Italian Pouzzolana, colour of wine dregs, containing	52.00	16.00	15.00	..	1.95	4.05	11.00
Artificial Pouzzolana, resulting from the calcination, in powder, of a brown ochreous clay, (1) containing	42.06	28.67	12.33	..	..	..	16.94
The same, from a plastic white clay, (2) containing	61.00	31.00	..	..	..	..	8.00
The same, from a grey, slightly effervescing clay, (3) containing	43.63	32.90	9.29	2.16	..	..	12.02
The same, from a brown ochreous clay, (4) containing	43.40	26.00	18.60	..	..	..	12.00
The same, from a red, very effervescent clay, (5) containing	42.60	15.96	8.00	24.64	..	..	8.80
ENERGETIC SUBSTANCES.							
Ochreous clay separated by washing from the arène of Perigord, containing	43.60	28.40	13.00	..	..	..	15.00
Artificial Pouzzolana, resulting from an ochreous clay, over-burnt, containing	42.06	28.67	12.33	..	..	..	16.94
SLIGHTLY ENERGETIC SUBSTANCES.							
Clay, separated from the arène of Montignac (Dordogne), containing	48.40	24.60	11.00	..	..	..	16.00
Clay, separated from the arene of Sancerre (Banks of the Loire), containing	47.60	22.00	10.40	..	..	..	20.00
Brown Psammite of Kergoat (Finisterre), of mediocre quality, containing	60.80	15.20	9.00	..	..	..	15.00
INERT SUBSTANCES.							
Quartzose sand	100.00	..	..	..	..	..	..
Ochreous clay heated to pasty fusion, containing	42.06	28.67	12.33	..	..	..	16.94
Scoriæ of large furnaces	37.00	19.00*	6.40†	..	38.60	..	..

* The Alumina contains from three to four hundredths of Magnesia.
† The Iron is in the state of protoxide.

No. III.

CHAPTER VII.

the behaviour of their ingredients in regard to Muriatic Acid.

Portions of these principles dissolved by digestion in Muriatic Acid ‡.				Behaviour of the Substances with rich Lime.				OBSERVATIONS.
				Proportions of the mixtures.		Solidification.		
Alumina.	Oxide of Iron.	Carbonate of Lime.	Principles not recognized.	Rich lime, measured in paste obtained by the first process.	Substances mixed with the lime in the pulverulent state.	Time of set.	Depression produced by the blow of the proving needle after a year's immersion.	
						Days.	Inches.	
12.20	5.60	..	9.40	2.00	1.00	8.00	0.0889	
9.00	11.40	..	5.60	{ 2.00 1.00	3.00 3.00	2.50 3.00	0.0728 0.0827 }	Splinter under the blow.
17.40	7.00	..	..	2.00	3.00	1.00	0.0787	Ditto.
12.40	..	..	..	2.00	3.00	2.50	0.0984	Ditto.
18.20	6.60	2·12	..	2.00	3.00	2.00	0.0866	Ditto.
15.00	9.80	..	..	2.00	3.00	2.50	0.0984	Ditto.
2.00	3.00	24·60	..	2.00	3.00	1.00	0.1181	
16.47	11.53	..	..	1.00	3.00	3.00	0.2405	The needle penetrates without splintering.
10.50	3.03	..	..	2.00	3.00	9.00	0.2657	Ditto.
7.93	5.55	..	..	1.00	3.00	18.00	0.3555	Ditto.
10.00	5.40	..	..	1.00	3.00	6.00	0.3228	Ditto.
5.20	8.80	..	..	1.00	3.00	17.00	0.4016	Ditto.
..	..	..	..	1.00	2.00	Never.	Soft.	
.. ..	.. ..	.. ..	.. ..	1.00 1.00	2.00 2.00	Never. Never.	Soft. Soft.	

‡ The Substances submitted to digestion in Muriatic Acid were in the state specified in the first column.

TABLE No. IV.

IN SUPPORT OF CHAPTER VIII.

COMPARISON OF VARIOUS ARTIFICIAL POUZZOLANAS with the Italian Pouzzolana, Audenack Tarras, and Aquafortis Cement.

DESCRIPTION OF THE INGREDIENTS.	PROPORTIONS OF THE CEMENTS PREPARED FROM THEM.		Time of set of the cement.	Depression produced by the blow of the proving needle, after a year's immersion.	OBSERVATIONS.
	Rich lime measured in paste obtained by the ordinary mode of slaking.	Ingredients in the state of powder.			
			Days.	Inches.	
Aquafortis cement of Paris	1.00	3.00	2.50	0.1102	All these cements splinter under the blow.
Italian pouzzolana, colour wine dregs red	1.00	3.00	3.00	0.0827	
	2.00	3.00	2.50	0.0728	
Audenack tarras	1.00	2.00	8.00	0.0889	
ARGILLACEOUS SUBSTANCES CALCINED IN POWDER EXPOSED TO THE AIR.					
Ochreous clay, (1) deposited by infiltration in the crevices of lime-stones, calcined in powder	1 00	3.00	0.50	0.0649	
	2.00	3.00	1.00	0.0787	
Ochreous clay, separated by washing from a brown arene (4) ditto	2.00	3.00	2.50	0.0984	
Effervescing clay from Baye (Nièvre) ditto	2.00	3.00	1.00	0.1181	
Plastic refractory clay of Loupiac (2) (Lot), ditto	2.00	3.00	2.50	0.0984	
Plastic clay of Nevers, ditto	1.00	3.00	1.50	0.1299	
Grey, slightly effervescing clay from Souillac (3) (Lot), ditto	2.00	3.00	2.00	0.0866	* This experiment, communicated by Mr. Avril, Engineer, was tried on a cement more than a year old.
Arene from Chatillon (Nièvre), ditto	1.00	3.00	1.50	0.1004	
Arene from Lamothe (Lot), ditto	2.00	3.00	1.50	0.1575	
Psammite from Kergoat	2.00	3.00	,, ,,	0.0492 *	
ARGILLACEOUS SUBSTANCES CALCINED IN POWDER IN CLOSE VESSELS.					† The cements furnished by these two experiments were sensibly deteriorated at their surfaces, to the depth of 0·1968 to 0·236 of an inch. They adhered to the sides of the glasses containing them with such force, that it was necessary to reduce them to powder to separate them.
Ochreous clay, cited (1)	2.00	3.00	5.00	0.0669 †	
Refractory clay, cited (2)	2.00	3.00	7.00	0.0787	
ARGILLACEOUS SUBSTANCES MODERATELY HEATED IN FRAGMENTS IN CONTACT WITH THE AIR.					The cements of these two series do not splinter under the blow.
Ochreous clay, cited (1)	1.00	2.00	4.00	0.1653	
	2.00	3.00	4.50	0.1889	
Pounded brick of the country, (Lot)	1.00	2.00	9.00	0.2476	
	2.00	3.00	12.00	0.2756	
Plastic clay, cited (2)	1.00	2.00	4.00	0.1811	
	2.00	3.00	5.00	0.2086	
Grey effervescing clay, cited (3)	1.00	2.00	5.50	0.1889	
	2.00	3.00	5.50	0.2559	
ARGILLACEOUS SUBSTANCES, AND OTHERS, POWERFULLY HEATED IN FRAGMENTS.					
Ochreous clay, cited (1)	2.00	3.00	,, ,,	0.4421	The commas indicate that the surface was never able to bear the needle which is used to measure the time of the first *set*.
Grey effervescing clay, cited (3)	2.00	3.00	,, ,,	0.6063	
Common slate	2.00	3.00	201.0	0.4819	
Ferruginous sand-stone	2.00	3.00	,, ,,	0.3756	
Basalt	2.00	3.00	,, ,,	0.6027	

NOTE.—When the calcination is pushed as high as the pasty state of fusion, the mortars become (with the exception of the slate) completely inert.

TABLE No. V.

IN SUPPORT OF CHAPTER IX.

COMPARISON INTENDED TO SHOW THE MUTUAL SUITABLENESS OF THE VARIOUS LIMES, with the different Ingredients of CEMENTS and MORTARS.

NAMES.	Proportions of the Mixtures.			Time of set of the immerged mortar or cement.	Depression produced by the blow of the proving needle after a year's immersion.
	Lime in paste obtained by the ordinary mode of slaking.	Quartzose sand.	Natural or artificial pouzzolana in powder.		
WATER MORTARS, Composed of eminently hydraulic lime, and quartzose sand, or inert substances.					
				Days.	Inches.
Eminently hydraulic lime of the valley of Rioges (Nièvre), with sand from the Loire	100	150	..	2.50	0.0866
The same, from Baye (Nièvre), with the sand of the Loire	100	150	..	4.00	0.1161
The same, from Vaux-Helles, near St. Privé, (Nièvre), and Loire sand	100	150	..	4.00	0.1338
The same, from the marl of Devay on the Loire, with Loire sand	100	150	..	4.00	0.0728
The same, from Fertotot, near Bec d'Allier, with Loire sand	100	150	..	8.00	0.0787
The same, from Quenou (Isle et Vilaine), with the quartzose sand of La Vilaine	100	150	..	4.00	0.1315
The same, from Pompeau (Isle et Vilaine), with the quartzose sand of La Vilaine	100	150	..	3.50	0.1531
The same from Vivier (Ardèche), with the granitic sand of Dordogne	100	150	..	2.00	0.0826
WATER CEMENTS, Composed of eminently rich lime, and very energetic pouzzolanas.					
Eminently rich lime, and Italian pouzzolana	100 100	.. ..	300 150	3.00 2.50	0.1181 0.1181
The same, with Audenack tarras	100	..	200	8.00	0.0889
The same, with aquafortis cement	100	..	300	2.50	0.1102
The same, with artificial pouzzolanas, obtained by calcining various clays, previously reduced to powder, mean result of 14 experiments given in Table No. 4.	100 200	.. ..	300 300	1.44	0.1015
WATER CEMENTS, Composed of moderately hydraulic limes, and simply energetic pouzzolanas, or mixtures of inert substances with very energetic pouzzolanas.					
Moderately hydraulic lime, and tile-dust; mean result of several experiments	100 100	.. ..	200 150	2.50	0.1354
The same, with well-burnt clays; mean result of many experiments	100 100	.. ..	200 150	3.00	0.1358
The same, with clays calcined in powder, tempered with one half of quartzose sand; mean of many experiments	100 100	.. ..	200 150	4.00	0.1149

TABLE

IN SUPPORT OF

Hydraulic Mortars and Cements Compared, with

NAMES.	Lime in paste.	Granitic Sand.	Tile Dust.	Artificial energetic pouzzolana.	Very energetic artificial pouzzolana.	Forge Scales.	Iron Dross.
MORTARS AND CEMENTS *Of eminently rich Lime from Laurac (Lot).* The mean depressions for the 1st, 2d, and 3d processes of slaking, respectively, are as follows : In. In. In. After the 1st year, 0.2888 0.1971 0.1802 After the 2d year, 0.1972 0.1578 0.1412	2.00	1.00	1.00	..	..	..	..
	1.33	..	2.00	..	..	..	..
	2.00	1.00	..	..	1.00	..	..
	1.33	..	..	..	2.00	..	..
	1.80	2.00	..	..	..	..	..
MORTARS AND CEMENTS *Of the eminently rich Lime from Loupiac (Lot).* The mean depressions for the 1st, 2d, and 3d processes of slaking, respectively, are as follows : In. In. In. After the 1st year, 0.3318 0.2066 0.1658 After the 2d year, 0.2784 0.1703 0.1370	4.50	1.00	..	1.00	..	..	..
	3.00	..	2.00	..	..	..	..
	4.50	..	1.00	..	..	..	1.00
	2.16	1.00	1.00	..	..	..	..
	2.16	1.00	..	..	1.00	..	..
	1.50	..	..	..	2.00	..	..
	1.50	1.00	..	..	..	1.00	..
	2.25	2.00	..	..	..	..	..
MORTARS AND CEMENTS *Of common rich Lime.* The mean depressions for the 1st, 2d, and 3d processes of slaking, respectively, are as follows : In. In. In. After the 1st year, 0.2390 0.1790 0.1563 After the 2d year, 0.2009 0.1402 0.1269	3.00	1.00	..	1.00	..	..	..
	2.00	..	2.00	..	..	..	..
	3.00	..	1.00	..	..	..	1.00
	2.16	1.00	1.00	..	..	..	..
	2.16	1.00	..	..	1.00	..	..
	1.44	..	..	..	2.00	..	..
	1.50	1.00	..	..	..	1.00	..
	1.35	2.00	..	..	..	..	..
MORTARS AND CEMENTS *Of feebly Hydraulic Lime.* The mean depressions for the 1st, 2d, and 3d processes of slaking, respectively, are as follows : In. In. In. After the 1st year, 0.2758 0.2019 0.1725 After the 2d year, 0.2407 0.1850 0.1514	3.00	1.00	..	1.00	..	..	..
	2.00	..	2.00	..	..	..	..
	3.00	..	1.00	..	..	..	1.00
	2 16	1.00	1.00	..	..	..	..
	2.16	1.00	..	..	1.00	..	..
	1.44	..	..	..	2.00	..	..
	1.50	1.00	..	..	..	1.00	..
	1.50	2.00	..	..	..	..	..
CEMENTS AND MORTARS *Of Hydraulic and eminently Hydraulic Lime.* The mean depressions for the 1st, 2d, and 3d processes of slaking, respectively, are as follows :	1.00	1.00	1.00	..	..	..	..
	1.00	..	2.00	..	..	..	..
	1.00	1.00	..	1.00	..	..	..
	1.00	..	..	2.00	..	..	..
	1.00	2.00	..	..	..	..	..
In. In. In. After the 1st year, 0.1263 0.1330 0.1969 After the 2d year, the experiments were not complete.	1.00	1.00	1.00	..	..	..	..
	1.00	..	2.00	..	..	..	..
	1.00	1.00	..	..	..	..	..
	1.00	..	..	..	..	..	..
	1.00	2.00	..	..	..	..	..

NOTE.—The cements and mortars not having been immersed in the same season, nor consequently, parison of the numbers which express the rapidity of set.

We purposely gave, varied, almost arbitrary proportions to the lime, in order to establish more last columns, indicate respectively, the use of the *Ordinary*, *Immersion*, and *Spontaneous*, modes of

No, VI.

CHAPTER X.

regard to the process of slaking made use of.

Time of Set.			Depressions produced by the blow of the proving needle.					
			After 1 year's immersion.			After 2 years' immersion.		
O	I	S	O	I	S	O	I	S
Days.	Days.	Days.	In.	In.	In.	In.	In.	In.
39.00	101.00	7.00	0.4421	0.2330	0.1964	0.2555	0.2252	0.1878
9.00	17.00	26.00	0.2476	0.1499	0.1610	0.1952	0.1204	0.1204
16.00	7.00	8.00	0 3004	0.2405	0.2027	0.2177	0.1653	0.1378
4.00	4.00	4.00	0.1653	0.1653	0.1610	0.1204	0.1204	0.1189
..	..	52.00	Soft	Soft	0.6527	Soft	Soft	0.6027
18.00	17.00	16.00	0.5311	0.1952	0.1653	0.4220	0.1728	0.1610
26.00	81.00	33.00	0.2102	0.1653	0.1653	0.1653	0.0976	0.0925
61.00	52.00	60.00	0.3614	0.1803	0.1610	0.2854	0.1653	0.1354
331.00	250.00	42.00	0.4752	0.3988	0.2177	0.3988	0.3161	0.1854
5.00	17.00	20.00	0.2634	0.1878	0.1728	0.2634	0.1653	0.1425
1.00	1.00	1.00	0.1499	0.1126	0.1126	0.1354	0.1051	0.1051
..	..	..	Soft	Soft	Soft	Soft	Soft	Soft
..	..	..	do.	do.	do.	do.	do.	0.6027
14.00	2.00	4.00	0.2779	0.2102	0.1610	0.2252	0.1952	0.1425
73.00	7.00	31.00	0.2177	0.1653	0.1610	0.1610	0.1204	0.1204
321.00	2.00	12.00	0.3082	0.1878	0.1653	0.2405	0.1051	0.0976
73.20	3.00	37.00	0.2476	0.2252	0.1803	0.2405	0.1425	0.1539
3.00	2.00	3.00	0.2177	0.1728	0.1653	0.2177	0.1653	0.1499
2.00	1.00	1.00	0.1653	0.1126	0.1051	0.1204	0.1126	0.0976
..	..	..	Soft	Soft	Soft	Soft	Soft	Soft
..	..	..	do.	do.	do.	do.	do.	do.
15.00	16.00	15.00	0.2712	0.2476	0.1728	0.2633	0.2252	0.1031
24.00	24.00	32.00	0.1610	0.1499	0.1499	0.1204	0.1499	0.1122
39.00	7.00	6.00	0.3311	0.1275	0.1248	0.2854	0.1204	0.1204
33.00	16.00	22.00	0.2405	0.1952	0.1653	0.2252	0.1499	0.1393
2.00	7.00	7.00	0.2027	0.1803	0.1425	0.1354	0.1354	0.1275
4.00	7.00	8.00	0.1126	0.0901	0.0826	0.1126	0.0752	0.0752
113.00	103.00	150.00	0.4456	0.3160	0.2712	0.4220	0.3161	0.2626
65.00	63.00	77.00	0.4421	0.3090	0.2712	0.3614	0.3082	0.2712
6.00	8.00	8.00	0.1315	0.1354	0.2283	0.1051	0.1161	0.1870
16.00	10.00	12.00	0.1511	0.1575	0.2401	0.1259	0.1299	0.2204
10.00	6.00	15.00	0.1338	0.1480	0.2440	0.1023	0.1023	0.2323
7.00	6.00	8.00	0.1618	0.1834	0.2598	0.1346	0.1220	0.2362
12.00	6.00	15.00	0.1220	0.1299	0.2161	0.1008	0.1023	0.1889
3.00	3.00	4.00	0.1122	0.1102	0.1515	..	..	..
4.00	3.00	6.00	0.1181	0.1220	0.1575	..	..	..
3.00	2.00	5.00	0.1144	0.1161	0.1535	..	..	..
3.00	2.00	4.00	0.1354	0.1417	0.1771	..	..	..
3.00	3.00	4.00	0.0826	0.0866	0.1417	..	..	..

in water of the same temperature, we ought not to attach much importance to the com-

clearly, the generality of the facts to be observed. The letters O, I, S, placed over the slaking.

TABLE No. VII.

IN SUPPORT OF CHAPTER X.

COMPARISON OF THE RELATIVE RESISTANCES OF VARIOUS HYDRAULIC MORTARS and CEMENTS, immerged in various States of Consistency.

NAMES.	Relative hardness measured very exactly by means of a piercer, after 8 months' immersion.
1st. A cement composed of 100 parts of the hydrate of rich lime slaked by immersion, and 200 parts of energetic artificial pouzzolana, having been immerged, of a good consistency, and submitted to proof by the piercer, after 8 months, offered a resistance represented by........................	1000
2d. The same, having been allowed to acquire a stiff consistency in the air, so that it was immersed quite in a bruised state, and without the least cohesion, gave under the same circumstances no more than	143
3d. The same, having been allowed to be altogether blanched, in very hot weather, so that it was immerged dry and pulverulent, without any cohesion, gave..	333
4th. The same, taken in the condition No 2, but bruised and beaten up again with additional water to the consistency No. 1, and immediately immerged, gave..	1000
5th. The same, taken in the condition No. 3, but bruised and beaten up with additional water to the consistency No. 1, and immediately immersed, gave	1000
6th. The same, taken in the consistency No. 2, but immerged with an envelope, and retaining the consistency it had already acquired, gave.........	2000
A second series of betons prepared with 100 parts of the hydrate of moderately hydraulic lime obtained by the ordinary extinction, with 50 parts of energetic pouzzolana, and 100 parts of quartzose sand, gave, under the same circumstances, and in the same order respectively, as below.	
Thus:	
1st..	1000
2d	116
3d	122
4th..	1000
5th..	1000
6th..	2000

Names	Consistency	Time of set.	Depressions produced by the blow of the proving-needle, after a year's immersion.
		Days.	Inches.
Mortar composed of 100 parts of the eminently hydraulic lime of the valley of Rioges (Nièvre), and 100 parts of the sand of the Loire, immerged ..	Of a firm and ductile consistency.	3.00	0.1181
	Of a very soft consistency........	7.00	0.2626
Mortar, ditto, of the eminently hydraulic lime of the neighbourhood of Champ Vert (Nièvre), and immerged..	Of a firm and ductile consistency .	3.00	0.1653
	Of a very soft consistency........	7.00	0.2992
Mortar, ditto, of the eminently hydraulic lime of Beauvoir (Nièvre), and immerged............	Of a firm and ductile consistency.	3.00	0.1515
	Of a very soft consistency........	5.00	0.2499
Mortar, ditto, of the eminently hydraulic lime of the domain of Fleury, dependence of Germancy (Nièvre), and immerged.....................	Of a firm and ductile consistency .	3.00	0.1212
	Of a very soft consistency........	5.00	0.3236

TABLE No. VIII.

IN SUPPORT OF CHAPTER X.

Hydraulic Mortars and Cements, compared with reference to the deterioration which they undergo at their surfaces.

COMPOSITION OF THE MORTARS AND CEMENTS.						Thickness of the deteriorated parts of various Hydraulic Mortars, two years old.			
Slaked lime in paste.		Granitic sand.	Forge scales.	Pounded brick.	Very energetic artificial pouzzolana.	Eminently hydraulic lime.	Moderately hydraulic lime.	Common rich lime.	Excessively rich lime.
By the ordinary process.	By immersion.					Inches.	Inches.	Inches.	Inches.
2.00	....	1.00	1.00	....	....	0.00	0.4330	0.5512	0.7086
....	2.00	1.00	1.00	....	....	0.00	0.3543	0.3937	0.6693
2.00	....	2.00	....	....	....	0.00	0.3543	0.4330	0.7086
....	2.00	2.00	....	....	....	0.00	0.2362	0.2756	0.7086
2.00	....	1.00	....	1.00	....	0.00	0.1968	0.2362	0.3543
....	2.00	1.00	....	1.00	....	0.00	0.1575	0.2362	0.3346
2.00	....	....	....	2.00	....	0.00	0.1575	0.2165	0.2756
....	2.00	....	....	2.00	....	0.00	0.1181	0.1181	0.2362
2.00	....	1.00	....	....	1.00	0.00	0.0590	0.1181	0.1968
....	2.00	1.00	....	....	1.00	0.00	0.0000	0.0984	0.1575
2.00	....	....	....	....	2.00	0.00	0.0000	0.0590	0.0787
....	2.00	....	....	....	2.00	0.00	0.0000	0.0000	0.0590

Quickness of Set of various Hydraulic Mortars and Cements, Compared with the resistance acquired after a year's immersion.

NAMES.	Time of set.	Depression produced by the blow of the proving needle.	Observations.
	Days.	Inches.	
Natural cement obtained from an argillaceous lime-stone at Belle-ville, on the banks of the Loire (Nièvre).	15.00	0.1889	
Hydrate of lime, prepared from a sandy lime-stone containing clay, calcined in contact with charcoal..	19.00	0.1771	
Cement, containing 100 parts of rich lime, and 300 parts of an artificial pouzzolana, prepared by slightly calcining a red effervescing clay, in fragments.	18.00	0.1220	
Mortar, composed of 100 parts of the eminently hydraulic lime of Chavance (Nièvre), and 150 parts of the sand of the Loire	11.00	0.1181	
Natural cement, prepared from an argillaceous lime-stone, found at Assigny, near Savigny (Nièvre)...	9.00	0.1153	
Mortar composed of 100 parts of eminently hydraulic lime of Fertotot (Nièvre), and 150 parts Loire sand.	8.00	0.0945	
Mortar, composed of 100 parts of eminently hydraulic lime, prepared from a sandy lime-stone containing clay, and 100 parts of Loire sand.	8.00	0.0787	
Hydrate of eminently hydraulic lime, very poor, obtained in the domain of Pont, near the rivulet of Creüll, on the banks of the Loire (Nièvre)	7.00	0.0433	
Mortar, composed of 100 parts of the eminently hydraulic lime of Vitry (Nièvre), and 150 parts of Loire sand.	8.00	0.3642	
Mortar, composed of 100 parts of the hydrate of moderately hydraulic lime, from Garnat, on the Loire, and 150 parts of Loire sand.	6.00	0.3366	
Cement, composed of 100 parts of the hydrate of rich lime, and 300 parts of the arène of Sancerre, on the banks of the Loire	6.00	0.3228	
Cement, of 100 parts hydrate of very rich lime, and 300 parts of the arène of Petreau (Gironde)	3.00	0.2886	
Cement, of 100 parts hydrate of very rich lime, and 300 parts of the arène of Herry, on the Loire.	6.00	0.2598	
Hydrate of hydraulic lime, obtained from the lime-stone of Chevanes, near Décise (Nièvre)	3.00	0.2145	

Note.—The object of these comparisons (which have been selected from experiments made with great care), is to show, that the time of set is not always an exact indication of the future hardness acquired by the immerged compounds.

TABLE No. IX.

IN SUPPORT OF CHAPTER X.

QUICKNESS OF SET OF VARIOUS CEMENTS, compared with the proportions, and the hardness acquired after a year's immersion.

NUMBER OF THE MORTARS.	Lime slaked by immersion, and measured in powder, not compressed.		Granitic sand.	Artificial pouzzolana.	Time of set.	Depressions produced by the blow of a steel point falling from a constant height.
	Very rich quality.	Moderately hydraulic.				
					Days.	Inches.
1	2.70	....	1.00	1.00	19.00	0.4752
2	2.00	....	1.00	1.00	16.00	0.2102
3	1.50	....	1.00	1.00	18.00	0.1988
4	1.00	....	1.00	1.00	8.00	0.1499
5	0.50	....	1.00	1.00	9.00	0.1653
1	2.70	....	....	2.00	8.00	0.1499
2	2.00	....	....	2.00	8.00	0.1228
3	1.50	....	....	2.00	7.00	0.0976
4	1.00	....	....	2.00	6.00	0.0901
5	0.50	....	....	2.00	6.00	0.1425
1	....	2.70	1.00	1.00	17.00	0.2405
2	....	2.00	1.00	1.00	16.00	0.1803
3	....	1.50	1.00	1.00	12.00	0.1803
4	....	1.00	1.00	1.00	8.00	0.1610
5	....	0.50	1.00	1.00	12.00	0.1913
1	....	2.70	....	2.00	12.00	0.1401
2	....	2.00	....	2.00	12.00	0.1370
3	....	1.50	....	2.00	10.00	0.1354
4	....	1.00	....	2.00	8.00	0.1275
5	....	0.50	....	2.00	10.00	0.1354

The object of this Table is to demonstrate, that the time of set is a sufficiently exact indication of the future hardness of immerged mortars and cements, when we confine our comparison to compounds containing the same elements, in various proportions.

TABLE No. X.

IN SUPPORT OF CHAPTER XI.

THE ABSOLUTE RESISTANCES OF MORTARS, compared with those of the HYDRATES OF LIME constituting their gangues.

	NAMES.	Lime in paste obtained by: Ordinary slaking.	Immersion.	Spontaneous.	Granitic sand.	Absolute resistances per square inch: Of the matrix, or lime.	Of the mixture, or mortar.
						lbs.	lbs.
MORTARS OF EMINENTLY HYDRAULIC LIME.	Mortar of eminently hydraulic lime from Labourgarde (Tarn et Garonne), composed of arbitrary proportions, 23 months old	....	1.14	....	1.50	70.45	123.8
		1.00	....	....	1.50	73.30	140.0
	Ditto, compounded in the proportions corresponding to the *maximum* resistance	1.00	....	....	1.80	73.30	165.40
	Mortar of eminently hydraulic lime from Montelimart (Drome), compounded in arbitrary proportions, one year old	1.00	....	....	1.50	96.64	126.54
	Mortar of eminently hydraulic lime from Vivier (Ardèche), compounded the same, 1 year old	1.00	....	....	1.50	82.55	101.91
	Mortar of eminently hydraulic lime from Baraigne (Lot), compounded in the proportions corresponding to the *maximum* resistance, and 14 months old	1.00	....	....	1.80	66.61	263.89
MORTARS OF MODERATELY AND FEEBLY HYDRAULIC LIME.	Mortar of moderately hydraulic lime from Saint Ceré (Lot), compounded in arbitrary proportions, and 1 year old	1.00	....	....	1.50	85.12	79.71
	Mortars of the feebly hydraulic lime of Cabessut at Cahors, prepared in arbitrary proportions, and 28 months old	1.00	....	....	1.50	108.75	34.16
	2d	....	1.00	....	1.50	59.21	53.23
	3d	....	....	1.00	1.50	44.12	66.61
MORTARS OF RICH AND VERY RICH LIME.	Mortars of the common rich lime of Souillac (Lot), compounded in arbitrary proportions, 22 months old	1.08	....	....	1.50	111.02	39.85
	Another	....	1.08	....	1.50	46.25	49.67
	Mortars of the very rich lime of Lanzac (Lot), composed of proportions corresponding with the *maximum* resistance, 6 years old	1.00	....	....	0.50	155.86	64.90
	Another	....	1.00	....	1.30	122.41	68.32
	Another	....	....	1.00	1.70	134.50	85.40

MORTARS compared with reference to size of the Sand made use of.

NAMES.	Proportions: Lime in paste.	Fine sand.	Coarse sand.	Small gravel.	Absolute resistances per square inch.
					lbs.
Mortars of the eminently hydraulic lime of Labourgarde (Tarn et Garonne), 23 months old	1.00	1.80	....	....	165.40
	1.00	....	1.80	....	134.50
	1.00	....	....	1.80	79.70
	1.00	0.90	0.90	....	148.02
Mortars of the eminently hydraulic lime of Baraigne (Lot), 14 months old	1.00	1.80	....	....	263.89
	1.00	....	1.80	....	177.20
	1.00	....	....	1.80	135.50
	1.00	0.90	0.90	....	222.04
Mortars of the moderately hydraulic lime of St. Ceré (Lot), 1 year old	1.00	1.80	....	....	84.68
	1.00	....	1.80	....	65.47
	1.00	....	....	1.80	48.68
	1.00	0.90	0.90	....	105.89
Mortars of the very rich lime of Lanzac (Lot), 22 months old	1.00	2.00	....	....	41.00
	1.00	....	2.00	....	54.66
	1.00	....	....	2.00	51.24
	1.00	1.00	1.00	....	47.82

NOTE.—It must be recollected, that each result is the mean of many experiments, made with all the exactness which this kind of investigation is capable of.

TABLE No. XI.

IN SUPPORT OF CHAPTER XI.

Mortars taken from various Buildings compared with Mortars manufactured for Experiment, with the same Limes.

NAMES.	Absolute resistance of the mortars per square inch.
Mortars of the eminently Hydraulic Lime of Montelimart. *Prepared by the ordinary process.*	
No. 1. Taken from a private dwelling-house in Montélimart, compounded with fine sand in good proportions, 19 years old.	lbs. 77.72
No. 2. Taken from an ancient tower at Montélimart, compounded like the above, 110 years old....................	104.19
No. 3. Manufactured at Souillac for experiment, with tolerably fine granitic sand, 12½ months old..........................	126.39
Mortars of the eminently Hydraulic Lime of Viviers. *Manufactured by the ordinary process.*	
No. 1. Taken from the wall of the rampart at Viviers, composed with moderately fine sand in good proportions, age unknown, but presumed to be not less than 600 years	140.77
No. 2. Prepared at Souillac for experiment, with moderately fine granitic sand, 12 months old	116.86
Mortars of the moderately Hydraulic Lime of Cahors. *Prepared by the ordinary process.*	
No. 1. Taken from the interior of a private mansion at Cahors. The mortar poor, the lime having been drowned; sand coarsish. Age 22 years	9.96
No. 2. The same, situated between two bricks; had undergone a very rapid desiccation ..	9.53
No. 3. The same, having dried in the ordinary way; better manufactured than the preceding	13.67
No. 4. The same, from the foundations of the same building, having undergone slow desiccation. Age 23 years	46.40
No. 5. Taken from the bridge of Valentré at Cahors, composed of coarse sand and gravel in good proportions, 400 years old	64.48
Mortars of the moderately Hydraulic Lime of Cahors—(continued). *Prepared by the ordinary process.*	
No. 6. Prepared at Souillac for experiment, with tolerably fine granitic sand, 20 months old ..	lbs. 34.16
No. 7. The same, with coarse sand and gravel, 22 months old	105.90
No. 8. The same, with coarse and fine sand mixed, 22 months old	91.00
Mortars of the very rich Lime of Lanzac (Lot). *Prepared by the ordinary process.*	
No. 1. Taken from a private house in Lanzac. The mortar from the cornice, rather poor than rich, compounded with tolerably fine granitic sand. Age 20 years...................	17.08
No. 2. The same, another fragment.........................	17.08
No. 3. The same, another fragment.........................	20.21
No. 4. The same, another fragment.........................	22.20
No. 5. Taken from a church, composed like the preceding, but poorer. Age 200 to 300 years.....................	33.02
No. 6. Prepared at Souillac for experiment, with the same sand, and 20 months old	32.02
Mortars of the very rich Lime of Loupiac (Lot). *Prepared by the ordinary process.*	
No. 1. Taken from a private mansion at Loupiac. The mortar rather poor than rich, composed with rather fine granitic sand, 17 years old....	21.77
No. 2. Taken from the church of Loupiac, rather meagre than rich, composed with pit sand. 200 years old	16.37
No. 3. The same, less meagre than the above	44.41
No. 4. Prepared at Souillac for experiment, with granitic sand. Age 20 months	18.07

TABLE No. XII.

IN SUPPORT OF CHAPTERS XI. AND XII.

MORTARS COMPARED in regard to THEIR PROPORTIONS, and the Process of Slaking made use of.

PROPORTIONS.		ABSOLUTE RESISTANCE OF THE MORTARS, PER SQUARE INCH.											
		Prepared from common rich lime, after 20 months.						Prepared from eminently hydraulic lime, after 14 months.					
Lime measured in paste.	Common granitic sand.	Exposed to the air.			Buried under ground.			Exposed to the air.			Buried under ground.		
		O	I	S	O	I	S	O	I	S	O	I	S
		lbs.	lbs.	lbs.	lbs.	lbs.	lbs.	lbs.	lbs.	lbs.	lbs.	lbs.	lbs.
100	..	103.19	54.09	64.05	..	..	..	66.04	59.50	44.55	149.17	141.63	125.26
100	50	44.41	34.16	41.28	2.70	3.13	3.56	90.53	85.40	79.71	161.13	153.72	131.66
100	60	37.01	35.58	41.28	3.27	2.84	4.27	95.65	92.52	78.28	163.97	148.60	137.21
100	70	34.16	32.74	38.43	2.42	3.70	3.56	120.13	108.46	83.98	162.83	149.45	131.66
100	80	41.28	40.56	42.70	2.70	3.56	6.12	157.71	148.02	95.65	164.54	155.15	133.80
100	90	42.70	42.70	44.41	2.27	3.42	4·41	169.67	156.57	117.43	169.81	141.63	136.64
100	100	37·01	46.40	49.53	2.13	3.27	4 27	174.22	169.38	153.72	170.80	162.26	138.07
100	110	35.58	46.40	42.70	2.27	2.56	4.41	179.35	162.26	149.45	171.37	161.13	145.18
100	120	39.85	55.51	48.39	2.42	2.70	6.40	184.47	173.65	145.18	175.93	161.70	146.61
100	130	41.28	58.07	51.24	2.27	2.84	4.27	191.87	179.35	165.11	175.08	153.01	148.02
100	140	41.28	66.33	58.07	2.27	3.42	4.13	211.80	196.43	162.42	174.22	160.84	157.99
100	150	39.85	63.76	58.07	2.42	4.41	4.13	239.13	202.83	175.36	177.64	177.50	148.02
100	160	41.28	67.61	63.76	2.27	3.70	5.41	248.24	236.28	199.27	174.51	156.57	135.93
100	170	44.41	68.32	64.90	2.42	3.70	5.41	256.21	256.21	213.51	171.37	156.57	130.52
100	180	44.12	75.15	68.32	2.13	3.42	4.98	263.75	253.36	210.66	170.80	172.23	159.42
100	190	45.55	81.13	81.98	1.71	5.41	5.12	261.62	241.98	193.58	167.67	163.69	145.18
100	200	48.39	78.28	78.57	1.14	5.12	5.41	240.84	238.42	172.23	170.80	162.55	146.61
100	210	47.82	81.13	81.98	1.71	5.12	5.55	222.62	219.20	162.26	174.22	160.84	135.22
100	220	51.24	81.98	95.08	1.56	4.41	5.41	213.51	202.83	159.42	177.64	142.34	136.64
100	230	51·24	73.16	85.40	1.56	5.12	5.41	204.97	172.23	150.88	189.17	145.89	131.66
100	240	51.24	75.15	63.76	1.56	3.42	5.55	198.14	176.50	142.34	179.63	146.61	132.37
100	250	49.82	65.47	71.74	1.56	4.41	5.41	....	....	....	....	....	....
100	260	47.68	68.32	63.76	1.42	6.83	7.54	....	....	....	....	....	....
100	270	41.28	69.74	63.76	1.28	6.83	7.26	....	....	....	....	....	....
100	280	41.28	64.05	62.63	1.14	4.41	6.12	....	....	....	....	....	....
100	290	49.53	55.51	59.78	1.28	2.70	4.13	....	....	....	....	....	....

NOTE.—The letters O, I, S, indicate that the lime used was prepared by the Ordinary mode of slaking, by Immersion, or by the Spontaneous method.

TABLE XIII.

IN SUPPORT OF CHAPTERS XI. AND XII.

MORTARS OF RICH LIME, FIFTEEN MONTHS OLD, compared in regard to the influence of manipulation.

PROPORTIONS.		Absolute resistance of the mortars per square inch.				OBSERVATIONS.
		Exposed to the air.		Buried under ground.		
Rich lime slaked by immersion, and measured in paste.	Common granitic sand.	Mixed in the ordinary manner.	Remixed with additional water during five months.	Mixed in the ordinary manner.	Remixed with additional water during five months.	
		lbs.	lbs.	lbs.	lbs.	
100	150	58.93	77.29	21.77	26.76	Each of these results is the mean of ten experiments, made with the greatest care, and differing little from one another.
100	200	81.13	89.96	26.33	28.47	The mortars remixed, were not operated on daily, but from time to time, at intervals of from 7 to 8 days. The mortars buried under ground were left in the open air for 2 months previous to being subjected to experiment.

MORTARS, compared with reference to the consistency given to the mixture of Lime and Sand.

PROPORTIONS.		Absolute resistance of the mortars at the age of 14 months, per square inch.						OBSERVATIONS.
		Exposed to the air.			Buried under ground.			
Eminently hydraulic lime, in paste obtained by the ordinary extinction.	Granitic sand.	Kneaded stiff.	Kneaded soft.	The same nearly liquid.	Kneaded stiff.	Kneaded soft.	The same nearly liquid.	
		lbs.	lbs.	lbs.	lbs.	lbs.	lbs.	
100	150	239.13	143.48	76.86	177.64	165.11	127.39	Each of these results is the mean of a number of experiments.
Rich lime 100	150	80.42	41.28	17.08	..	..	..	

MORTARS, compared in regard to the influence of desiccation.

PROPORTIONS.		Absolute resistance of the mortars when 22 months old, per square inch.		OBSERVATIONS.
Lime slaked by immersion, and measured in paste.	Common granitic sand.	When exposed after manufacture, in a loft of the temperature of 59° Farenheit *.	At first put under ground, and afterwards removed into the open air by degrees.	
		lbs.	lbs.	
Hydraulic lime 190	150	92.23	165.68	Each of these results is the mean of a number of experiments.
Feebly hydraulic lime 180	150	44.12	73.73	
Common rich lime 180	150	45.55	51.24	

* The temperature stated in the original is 15°, which I have reduced on the supposition that it is meant to refer to the Centigrade thermometer; fifteen degrees of Reaumur's scale would correspond to 66° of Fahrenheit, very nearly.—TR.

TABLE No. XIV.

IN SUPPORT OF CHAPTER XIV.

The absolute Resistance of various Mortars, compared in regard to the effect of condensation.

NAMES.	Rich Lime in paste.		Hydraulic Lime.	Ingredients.					Absolute resistance of the mortars, per square inch, at the age of 12 months.		
				Sands.		Powders.					
	Obtained by the spontaneous extinction.	Obtained by the ordinary extinction.	In paste, obtained by immersion.	Common granitic sand.	Very fine granitic sand.	Chalk.	Building limestone capable of being cut in slabs	Crystallized carbonate of lime.	Prisms not compressed.	Prisms compressed in the direction of their length.	Prisms compressed at right angles to their length.
									lbs.	lbs.	lbs.
Mortars exposed to the weather......	100	..	..	220	..	..	..	..	17.79	31.31	32.74
	..	100	..	220	..	..	..	..	9.82	19.78	37.57
	..	..	100	180	..	..	..	..	165.54	119.56	170.81
	..	..	100	..	180	..	..	..	133.23	129.24	232.29
	..	..	100	..	..	180	..	..	122.98	99.64	143.48
	..	..	100	..	..	..	180	..	187.88	88.25	218.63
	..	..	100	..	..	..	..	180	191.45	136.64	225.46
	..	..	100	..	..	..	..	..	44.41	47.82	93.94
Mortars buried under ground.......	100	..	..	270	..	..	..	..	10.82	26.19	13.66
	..	100	..	270	..	..	..	..	8.11	23.20	21.92
	..	..	100	150	..	..	..	..	102.48	157.14	133.23
	..	..	100	..	..	180	..	..	6.83	68.46	54.65
	..	..	100	..	..	..	180	..	71.74	88.82	88.11
	..	..	100	..	..	..	..	180	59.21	102.48	95.65
	..	..	100	..	..	..	..	..	122.41	65.19	89·67

Various Mortars and Cements, compared in regard to their specific gravity, and porosity.

NAMES.	Specific gravity.	Weight of the water absorbed by imbibition, per cubic foot.
		lbs.
Various mortars of hydraulic lime, and quartzose, or calcareous sand, subjected to condensation......	2.031	10.99
	2.000	10.74
	1.960	12.12
	1.945	12.68
	1.928	12.24
Various mortars of hydraulic lime, and sand, or calcareous powders, not compressed, but kneaded stiff......	1.885	12.24
	1.873	13.55
	1.854	12.18
	1.789	14.93
	1.753	15.68
	1.747	15.55
Various mortars of hydraulic lime, and quartzose or calcareous sands, mixed thin	1.692	17.12
	1.692	17.55
	1.317	23.36
	1.279	22.99
	1.267	24.30
Various cements taken from ancient aqueducts......	2.000	9.74
	1.940	19.18
	1.760	19.99
	1.602	18.80
	1.516	22.05
	1.472	21.55
	1.402	24.99
	1.384	20.93
	1.359	18.99
	1.342	23.99

Note.—It results from these comparisons, that the cements are in general more permeable, than well manufactured mortars of lime and sand.

TABLE No. XV.

IN SUPPORT OF CHAPTER XVI.

Characters, Composition, and absolute Resistances of some Roman Mortars from the South of France.

NAMES.	Composition of the Mortars.		Observations.	Absolute resistances per square inch.
	Substances reduced to impalpable powder, and forming the principal matrix.	Palpable substances imbedded in the matrix.		No. lbs.
MORTARS FROM LARGE OR MASSIVE MASONRY.				
FROM THE ANCIENT VESUNA. No. 1. Taken from the Amphitheatre. 2. Taken from the Baths. 3. Taken from the Tower of Vésuna.	The matrix red; formed of white lime, and a very small quantity of red pouzzolana in impalpable powder.	Quartzose gravel, from the size of a pea to that of a walnut.	The mortar rich, coarsely amalgamated, in it a multitude of lumps of lime were visible.	1. .. 35.30 2. .. 76.01 3. .. 95.37
FROM CAHORS. No. 1. Taken from an ancient theatre. 1'. The same. 2. From a temple. 3. From a Roman canal.	Matrix of a dirty white, formed of a lime of the same colour.	Quartzose gravel, from the size of a pin's head, to that of a hazel nut.	The mortar rich, better worked than the preceding, containing no pouzzolana.	1. .. 60.35 1'... 64.90 2. .. 53.94 3. ..109.32
FROM NISMES. No. 1. Taken from the Amphitheatre. 2. From the Temple of Diana. 3. From the Tour Magne (Turris Magna).	The matrix sometimes grey, sometimes red, formed of a white lime, and of a small quantity of red pouzzolana in impalpable powder.	Quartzose and calcareous gravel, from the size of a pea, to that of a biggish hazel nut.	The mortar rich and very badly mixed, a multitude of lumps of white lime were visible in it.	1. .. 52.52 2. .. 79.56 3. ..169.38
FROM THE AQUEDUCT OF GARD. No. 1. Taken from the masonry. 1'. Ditto.	Matrix earthy, prepared from lime and a terreous sand.	Quartzose gravel, of the pretty nearly uniform size of a pea.	The mortar poor, ill-mixed, full of lumps of white lime.	1. .. 87.82 1'... 45.69
FROM VIENNE. No. 1. Taken from the Amphitheatre. 2. From the drains ("egouts").	Matrix tolerably homogeneous, of a dirty white, composed of lime only.	Quartzose gravel of every dimension, from that of a pea, to a good sized hazel nut.	The mortars rich, tolerably well mixed.	1. .. 82.55 2. .. 25.62
FROM UXELLODUNUM. (*At Capdenac on the Lot.*) No. 1. Taken from thick masonry.	Matrix grey, formed from a grey lime only.	Quartzose gravel of every dimension.	The mortar rich, badly worked, full of lumps of lime.	1. .. 51.95
MORTARS FROM ANCIENT AQUEDUCTS, REVETMENTS, FLOORS OR PAVEMENTS.				
FROM THE ANCIENT VESUNA. No. 1. Taken from the aqueduct of the baths.	The matrix reddish, formed of a white lime, and red brickdust in impalpable powder.	Fragments of red brick of the size of a walnut, in small quantity.	The mortar very rich, badly mixed, filled with a multitude of lumps of white lime.	1. .. 63.20
FROM CAHORS. No. 1. Taken from the aqueduct of the baths. 2. From the plastering of an ancient gate. 3. From a basin.	Matrix reddish, composed of a white lime, and red brick in impalpable powder.	A small quantity of fragments of red brick, of the size of a hazel nut.	The mortar rich, mixed in a middling degree, filled with portions of lime of a dirty white.	1. ..115.86 2. .. 29.04 3. .. 98.64
FROM NISMES. No. 1. Taken from the galleries of the Amphitheatre.	The matrix grey, composed of a white lime, and charcoal powder.	Fragments of charcoal as big as a pea.	The mortar tolerably well made.	1. ..127.11
FROM THE AQUEDUCT OF GARD. No. 1. Taken from the revetment of the aqueduct. 1'. The same.	The matrix whitish, composed of lime only.	Fragments of yellow and red brick of the size of a walnut.	The mortar very rich.	1. .. 43.98 1'...118.85
FROM VIENNE. No. 1. Taken from an ancient mansion —Revetment. 2. Ditto pavement. 3. Plastering.	Matrix reddish, composed of lime, and red brick in impalpable powder.	Fragments of red brick as big as a hazel nut.	The mortars tolerably well amalgamated.	1. .. 55.08 2. .. 78.85 3. ..102.48
FROM SARLAT. No. 1. Taken from a Roman aqueduct.	The matrix reddish, composed of lime, and red brick in impalpable powder.	Fragments of red brick as big as a walnut.	The mortar tolerably well mixed.	1. .. 55.08

TABLE No. XVI.

IN SUPPORT OF CHAPTER XVII.

COMPARISON of the RESISTANCES of VARIOUS COMPOUNDS, in regard to Proportions, and the Size of the Bodies imbedded in the Substance constituting the Matrix.

Number of the bricks.	Common sand.	Coarse sand.	Gravel.	Clay.	Absolute resistance of the bricks, per squareinch.	OBSERVATIONS.
					lbs. aver.	
1	...	..	..	1.00	58.07	
2	1.00	..	..	1.16	18.93	
3	1.00	..	..	0.96	14.94	
4	1.00	..	..	0.76	12.95	Bricks Nos. 12, 13, 14, 15, and 16, were incapable of supporting the very light box into which the sand is poured.
5	1.00	..	..	0.56	11.67	
6	1.00	..	..	0.36	8.96	
7	....	1.00	..	1.16	4.55	
8	....	1.00	..	0.96	3.56	
9	....	1.00	..	0.76	3.27	
10	....	1.00	..	0.56	0.43	
11	....	1.00	..	0.36	0.00	
12	....	..	1.00	1.16	....	
13	....	..	1.00	0.96	....	
14	....	..	1.00	0.76	....	
15	....	..	1.00	0.56	....	
16	....	..	1.00	0.36	....	
				Plaster of Paris.		
1	....	..	..	1.00	209.10	
2	1.50	..	..	2.00	120.99	
3	1.50	..	..	1.00	84.97	
4	....	1.50	..	2.00	108.89	
5	....	1.50	..	1.00	59.64	
6	....	..	1.50	2.00	88.82	
7	....	..	1.50	1.00	39.57	

TABLE No. XVII.—(Tr.)

REFERRED TO IN NOTES, TO ARTICLE 309, AND APPENDIX XXXI.

Analyses of various old Mortars.

Number.	Names.	Specific gravity.	Age.	Lime.	Carbonic acid.	Silica.	Other insoluble matter.	Magnesia.	Alumina.	Water and loss.	Total analysed.	Observations.
			Years.									
1.	From a Dutch tomb at Masulipatam ..	1·67	120	53.2	38.5	101.0	..	..	..	7.3	200	
2.	Ditto, ditto, another specimen	..	120	76.0	56.0	160.0	..	..	..	8.0	300	
3.	Ditto, ditto, another	..	120	26.0	17.0	153.0	..	..	..	4.0	200	
4.	From a Mahometan tomb............	..	200	115.9	85.0	188.0	..	..	..	11.0	400	
5.	From a tomb on the road from Masulipatam to Hyderabad..............	2.13	200	99.7	73.0	171.2	2.5	..	..	3.6	350	The insoluble matter is of a vegetable nature. (Nos. 5–7)
6.	Another specimen from the same....	..	200	110.9	80.0	199.3	2.0	..	..	7.8	400	
7.	Another do. from do.	1.90	200	110.4	79.0	206.5	0.5	..	..	3.6	400	
8.	Another do. from do.	2.13	200	19.7	12.0	64.5	..	..	..	3.8	100	
9.	From another tomb..	2.12	150	16.4	10.2	30.0	..	..	..	3.4	60	
10.	From the same	..	150	17.0	10.3	50.0	..	..	..	2.7	80	
11.	From the Pagoda at Tripatty.........	1.5	400	17.4	11.9	60.0	..	9.52	..	1.18	100	
12.	Ditto, ditto, another specimen	1.48	400	13.8	8.0	72.0	..	5.6	1.8	..	100	Excess = 1.2
13.	From a tomb at Golcondah, near Hyderabad	..	200	33.9	20.0	10.0	..	..	..	11.1	75	
14.	From a Roman tower near Boulogne....	..	1800	29.1	20.0	45.0	..	..	..	5.9	100	
15.	Decayed plaster	..	recent.	17.1	10.0	65.0	..	..	..	6.9	100	These specimens were quite in a disaggregated condition. (Nos. 15–16)
16.	Ditto, ditto, another	..	ditto.	15.6	10.0	70.0	..	..	..	4.4	100	
17.	Plastering from a public building in Masulipatam	1.75	45	60.1	39.5	292.5	3.0	..	..	4.9	400	The insoluble matter is vegetable.

INDEX.

A.

D.

E.

F.

M.

P.

S.

W.

Y.

G. Woodfall, Printer, Angel Court, Skinner Street, London.

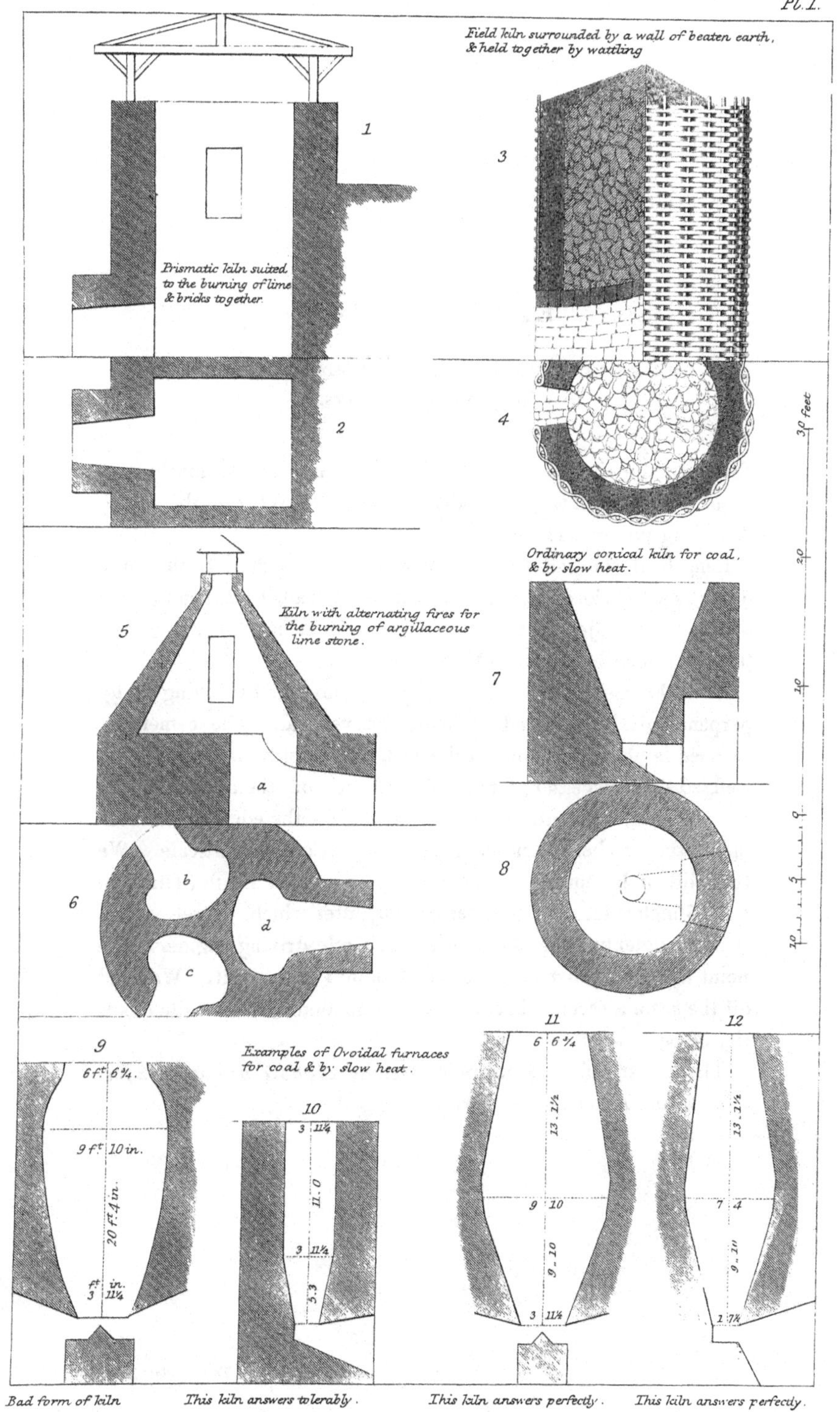

J. Hawkesworth sc.

London, Published May, 1837, by John Weale, 59, High Holborn.

EXPLANATION

OF THE MACHINE FOR TRYING THE HARDNESS OF HYDRAULIC MORTARS OR CEMENTS.

THE average dimension of the different parts of the machine is about an inch each way; breadth 10 inches, height from the sole to the lifting pulley, 21½ inches.

Length of the rod *a b*, viz.: from *a* to *c*, 6 inches, from *c* to *d*, for the part enclosed within the cylinder of lead, 1.69 inches, from *d* to *b*, to the adjustment of the point *p*, 5½ inches; section ¼ inch (nearly); length of lift 3.937 inches.

To make use of this machine, we commence by setting it up perpendicularly; the rod *a b* being kept vertical. The cement to be tried is placed underneath this rod, the vessel containing it being wedged up if necessary: the point of the rod then bears on the surface of the cement. We read off at *e*, on the edge of the bent index, the number of tenths of an inch marked on the scale. We then lift it by means of the string *f* to a given height, (fixed at 1.9685 inches for all our experiments,) after which, we release the string suddenly, like the monkey of a pile-driving engine. The point falls and penetrates more or less into the cement. We read off the scale a second time, and by subtraction arrive at the quantity of *penetration*.

The rod armed with its point weighs 15383.7 grains, or 2 lbs. 3 oz. 2½ dr. Avoirdupois nearly.

To face Plate II.

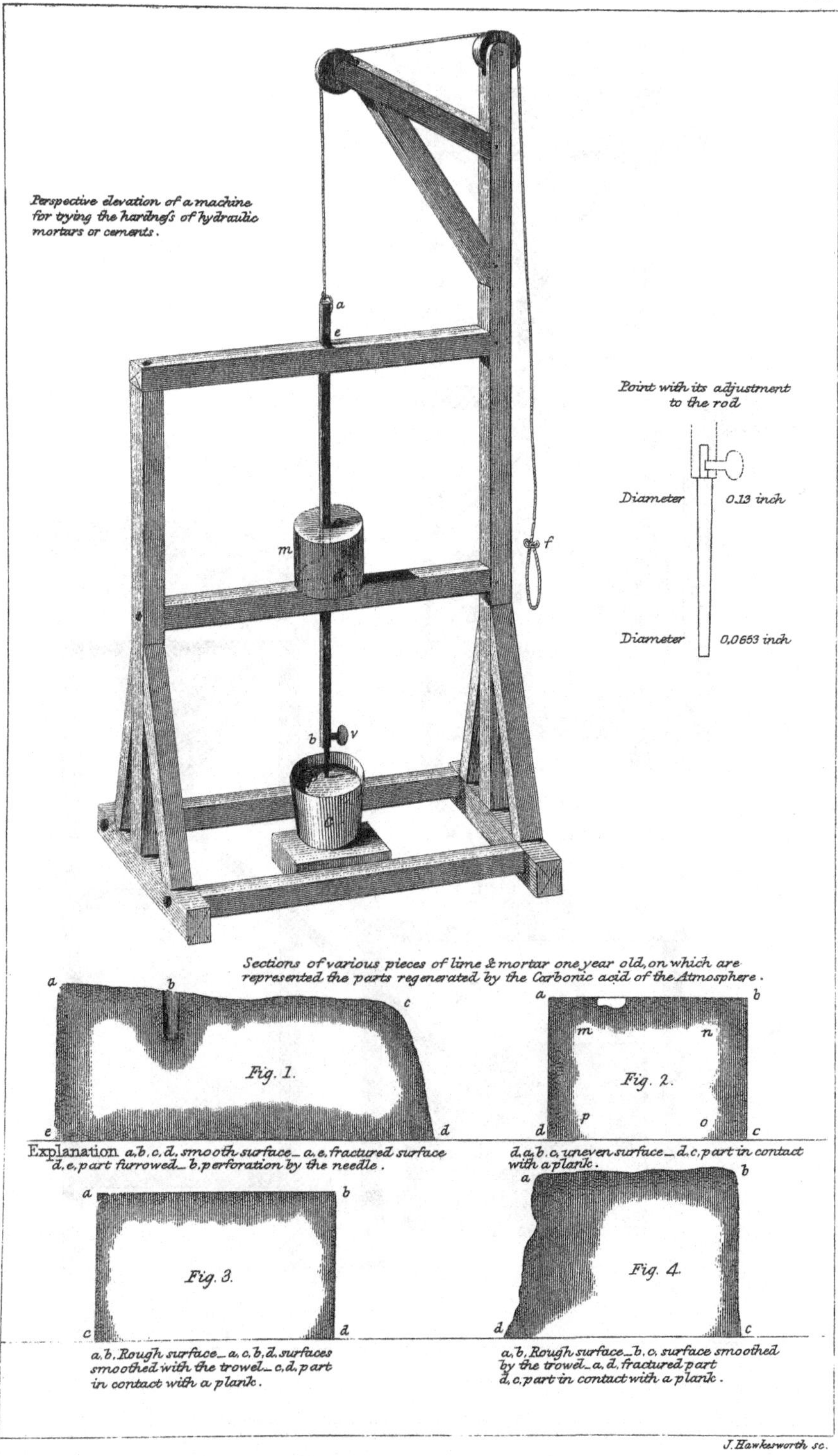
Perspective elevation of a machine for trying the hardness of hydraulic mortars or cements.
a
e
m
f
b
v
c
Point with its adjustment to the rod
Diameter 0.13 inch
Diameter 0,0653 inch
Sections of various pieces of lime & mortar one year old, on which are represented the parts regenerated by the Carbonic acid of the Atmosphere.
Fig. 1.
a b c d e
Explanation a, b, c, d, smooth surface _ a, e, fractured surface d, e, part furrowed _ b, perforation by the needle.
Fig. 2.
a b m n p o d c
d, a, b, c, uneven surface _ d, c, part in contact with a plank.
Fig. 3.
a b c d
a, b, Rough surface _ a, c, b, d, surfaces smoothed with the trowel _ c, d, part in contact with a plank.
Fig. 4.
a b d c
a, b, Rough surface _ b, c, surface smoothed by the trowel _ a, d, fractured part d, c, part in contact with a plank.
J. Hawkesworth sc.

London, Published May, 1837, by John Weale 59, High Holborn.

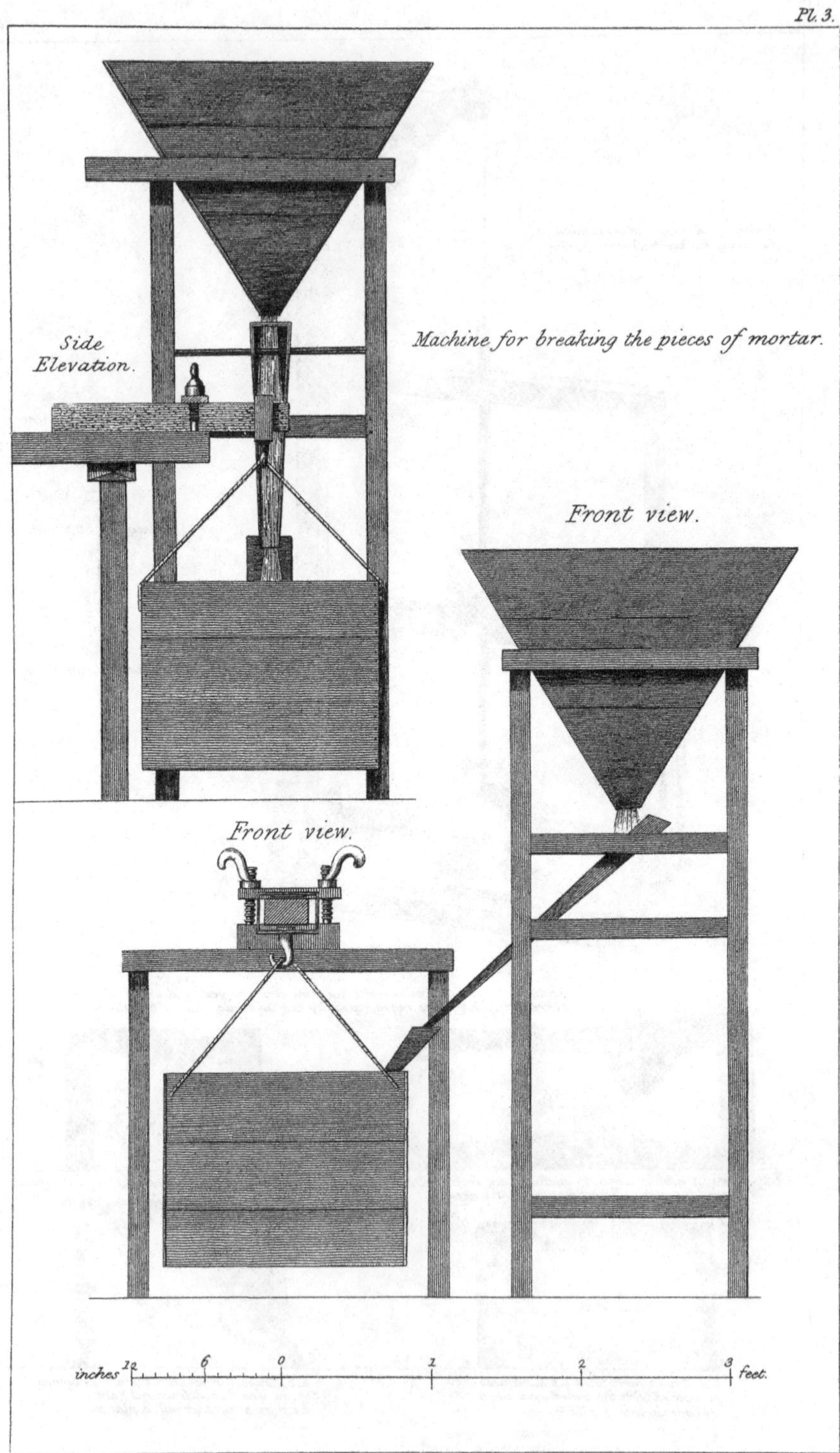

J. Hawkesworth sc.

London, Published May, 1837, by John Weale. 59. High Holborn.

Printed in the United States
By Bookmasters